Technical Writing

Second Edition

101

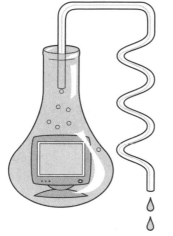

A Real-World Guide to Planning and Writing Technical Documentation

by
Alan S. Pringle
Sarah S. O'Keefe

SCRIPTORIUM
press

Technical Writing 101: A Real-World Guide to Planning and Writing Technical Documentation

Second edition

Copyright © 2000–2003 Scriptorium Publishing Services, Inc.

Scriptorium Press, Scriptorium Publishing Services, and their logos are trademarks of Scriptorium Publishing Services, Inc. All other trademarks used herein are the properties of their respective owners and are used for identification purposes only.

No part of this book may be reproduced or transmitted in any form or by any means, graphic, electronic, or mechanical, including photocopying, recording, taping, or by any information storage or retrieval system, without written permission from Scriptorium Publishing Services, Inc.

Chapter 14 is adapted from Scriptorium Publishing's white paper, "Structured Authoring and XML," which was originally published on the World Wide Web at http://www.scriptorium.com/structure.pdf.

Published by Scriptorium Press, the imprint of Scriptorium Publishing Services, Inc.

For information, contact:
Scriptorium Publishing Services, Inc.
PO Box 12761
Research Triangle Park, NC 27709-2761 USA
Attn: Scriptorium Press
www.scriptorium.com/books
books@scriptorium.com

ISBN: 0-9704733-2-X

Printed in the United States of America

Second printing

About the authors

Alan S. Pringle is lead technical editor at Scriptorium Publishing Services, Inc., and director of its publishing imprint, Scriptorium Press. The company provides technical publishing services to high-tech companies, including outsourced documentation solutions, technical training, and consulting. Since 1990, he has worked as a technical writer and editor on projects ranging from writing user guides for laser printers to editing course material for telecommunications equipment. Alan has also established corporate style guidelines and implemented XML-based structured applications for FrameMaker workflows. He is coauthor of the *FrameMaker 7 Workbook Series.* Alan lives in Cary, North Carolina. He enjoys finding bargains at flea markets, thrift stores, and auctions, and he likes all types of movies (even some really bad ones).

Sarah S. O'Keefe is founder and president of Scriptorium Publishing Services, Inc. Sarah is an experienced FrameMaker trainer; she has both Certified Technical Trainer (CTT) and FrameMaker Adobe Certified Expert (ACE) credentials. Her background also includes technical writing, technical editing, production editing, and extensive online help development with various help authoring tools. Sarah is coauthor of *Publishing Fundamentals: FrameMaker 7.* Currently, she works as a consultant to assist companies in implementing publishing solutions, including XML-based structured authoring. In her spare time, she likes to cook.

Acknowledgments

Special thanks to the following people, who helped make this a better book:

- Karen Brown for her work on the index
- Bill Burns for writing the chapter on localization and internationalization
- Sean Byrne for his illustrations
- Dave Kmiec for editing the book and drawing some graphics
- Sheila Loring for her contributions to the production editing chapter and for completing the production edit

Alan Pringle would like to dedicate this book in memory of his mother.

Contents

Preface

For technical writers, answering the obligatory "What do you do for a living?" question at a party can have many effects. It can:

- Create more questions: *"What's that?"*
- Draw blank stares
- Provoke some minor hostilities: *"Did you write that worthless manual that came with my word processing software? That book reeked!"*[1]

So, if you've decided that you want a career in technical communication, be prepared. Although you'll have a challenging, fast-paced job that changes as swiftly as the technology you write about, discussing your work at a party will be quite a conversation stopper.

What's in this book

Technical Writing 101 will show you that there's more to technical writing than just writing. The first major section of the book explains the skills you need as a writer. It also describes some of the essential tools and

---

1. Notice the skillful use of bullets, the hallmark of a good technical writer.

techniques for delivering projects on schedule and on budget. The chapters in this section are:

- Chapter 1, "So, what's a technical writer?"

  Explains what technical writers do and what skills they need.

- Chapter 2, "An overview of the technical writing process"

  Provides a high-level view of the technical writing process. Most documentation projects share a common structure—even when the subject matter is completely different.

- Chapter 3, "Very necessary evils—doc plans and outlines"

  Explains two project management tools that every technical writer needs. Documentation plans provide writers with a roadmap to follow as they create materials. Typically, a documentation plan includes a description of the target audience, the schedule, and a list of documents or online help to be developed. Manual outlines are just that—those hideous indented things you probably remember from high school. Unfortunately, you'll discover that outlines are in fact a necessity for technical writers, but perhaps you'll be convinced that they aren't so bad when you're trying to write an entire book!

- Chapter 4, "The Tech Writer's Toolbox"

  Focuses on the tools and technologies that you need to work successfully. Technical writers use a variety of writing and graphics packages to develop material.

In the second major section, you learn about how to get information, organize information, and (finally) write documentation. The chapters also describe other tasks in the documentation process, such as creating graphics, technical editing, production editing, and indexing. The chapters in this section are:

- Chapter 5, "Getting information"

  Gives you tips on how to extract information from source documents and product developers. Many people respond well to bribery, especially when the bribe is edible and includes chocolate in some form.

- Chapter 6, "Finally—it's time to start writing"

  Describes how to write documentation. When writing, it's important to address the document's audience and to divide your content into different types of information—interface, reference, conceptual, and procedural.

- Chapter 7, "Writing task-oriented information"

  Explains the basics of writing procedures. Because most technical documents tell users how to perform tasks, the ability to write good task-oriented information is a fundamental skill for all technical writers.

- Chapter 8, "A few words about pictures"

  Describes how to create and work with images. A graphic may not be worth exactly a thousand words, but an illustration can often explain something with more clarity than any amount of text.

- Chapter 9, "Editors—resistance is futile"

  Explains what you can expect from an editor and what most editors can expect from you. Refusing to work with an editor is not an option for technical writers—editing is an essential component of the technical documentation process. A competent editor can make you look good by catching your mistakes before the client sees them.

- Chapter 10, "Indexing"

  Explains the basics of writing a good index. A thorough, useful index is essential because readers often check the index first when looking for a particular piece of information. A good index can also save a company money—readers who quickly find the information they need are less likely to call customer support.

- Chapter 11, "Final preparation—production editing"

  Tells you how to make sure your document is ready for printing by checking for line breaks, page breaks, and other formatting issues.

The third section explains some advanced topics. Chapters in this section are:

- Chapter 12, "Avoiding international irritation"

  Offers some tips on minimizing the hassles that occur when documentation is translated into other languages. Learning about the translation process before you start writing the English documentation can save your company a lot of time and money—and prevent many, many headaches.

- Chapter 13, "Single sourcing"

  Describes how to create multiple types of deliverables—hardcopy books and online help, for example—from one set of files. The ability to create multiple deliverables while minimizing the time and money spent is important for many documentation departments, which often operate under tight schedules and with limited budgets.

- Chapter 14, "Structured authoring with XML—the next big thing"

  Explains the new trend of implementing structured authoring—a publishing workflow that defines and enforces consistent organization of information in documents—and how it affects technical writers.

The appendices provide information about getting a job, lists of resources and tools, and a sample documentation plan. The appendices are:

- Appendix A, "Getting your first job as a technical writer"

  Gives you some pointers on tailoring your resume for technical writing jobs, interviewing, and putting together a portfolio.

- Appendix B, "Resources"

  Lists web sites, books, and organizations that are useful for technical writers.

- Appendix C, "Tools information"

  Lists tools technical writers use.

- Appendix D, "Sample doc plan"

  Contains a sample documentation plan.

This book focuses on documentation for computer hardware and software. However, many of the concepts described apply to other forms of technical writing, such as writing about manufacturing environments, medical and pharmaceutical topics, and science.

If you're a talented writer with an interest in technical topics, writing technical documentation can be quite lucrative. This book gives you the advice and tools you'll need to get started in this challenging field.

1 So, what's a technical writer?

What's in this chapter

- ❖ Knowledge of technology
- ❖ Writing ability
- ❖ Organizational skills
- ❖ Strong detective (and people) skills

The short definition of "technical writer" is a person who writes about technical topics. But perhaps a better definition is someone who can explain complicated concepts in clear, easy-to-understand prose.

A technical writer is really a translator. You start with a complicated piece of technology, and your mission is to explain to a nonexpert how to use that technology.

This deceptively simple mission requires more than just writing ability and an understanding of technology— although both of those skills are critical, they aren't enough.

As a technical writer, you need strong organizational skills because you have to organize the information you gather. In many cases, getting information isn't difficult, but identifying what's relevant for your readers can be challenging. Many experienced technical writers create several new books each year. They must absorb new information and then understand, organize, prioritize, and deliver it.

You also need people skills and investigative talent. The stereotypical crabby, difficult starving artist/writer won't make it as a technical writer. The information you need resides in someone's head, so you must be able to work with that person to (gently?) extract the information.

This chapter describes the four basic skill sets every technical writer needs:

- Knowledge of technology
- Writing ability
- Organizational skills
- Detective and people skills

You'll find technical writers who are stronger in one skill set and weaker in another, but every successful technical writer needs to be competent in these four basic areas.

Knowledge of technology

The "technical" part of technical writing doesn't necessarily mean that you must be a programmer, electrical engineer, or other hardcore techie. However, you should be comfortable with and have some basic knowledge about the technology you'll be documenting. For example, if you plan to become a software technical writer, you should understand how to use an operating system (such as Windows), and you should be comfortable using various applications.

You should also understand basic concepts such as databases, networking, the Internet, and file manipulation (opening, closing, and saving). You should be willing (or even eager) to learn about new technology as it develops. Remember, you're the one who's expected to write the documentation, so you need to be able to figure out how the hardware or software works.

That doesn't mean, though, that you're completely on your own. In most jobs, you have access to information about the product you're documenting and can ask the programmers or engineers for details.

If your job requires that you write about something particularly technical, you need a higher level of knowledge. For example, if you're documenting a C++ programming tool, you should have a basic grasp of C++ programming. You don't have to be a programmer, but you do need to understand the fundamentals of programming.

You can learn about new developments in technology by subscribing to computer magazines or online mailing lists, or by joining user groups in your community. See Appendix B for more information about resources for technical writers.

Typing 101

If you're a good typist, you may be tempted to put your typing speed on your resume. Don't. Decent typing speed is a basic necessity for any technical writer, much like breathing. By putting typing speed on your resume, you are implying that you will be evaluated only on how fast you can pound out words. And that's not the case—you need to pound out the *right* words.

Your typing technique is not important. Touch typing is nice, but hunting and pecking is also fine, provided that you can type at about 40–50 words per minute. If using a keyboard is a laborious, error-prone, and tedious task for you, consider getting some typing software and improving your typing speed.

If spending most of the day at a keyboard typing sounds unappealing, technical writing is probably not the job for you.

Ignorance is bliss

Writing about new technology (often nonexistent at the beginning of the project—and sometimes surprisingly close to the end) means that you can't possibly be expected to know everything about the new stuff. Your initial lack of knowledge, however, works for you. As you learn how to use the new hardware or software, you encounter many of the same issues that the users (people who buy the product) will face as they get started.

This experience should play a critical part in how you write your documentation: "I had a hard time figuring out how to use the interface to edit the file, so I need to be sure I explain that thoroughly to new users."

Initial ignorance—or, even better, the ability to pretend ignorance—can be a valuable asset when you're writing documentation. But don't use this research technique to avoid learning about existing technology. By understanding what's already out there, you build a strong foundation for writing about what's to come.

Who treats the doctor and who documents for the writer?

As a technical writer, you need to be able to figure out new concepts with little or no assistance. Often (always?), product developers work long hours to meet aggressive deadlines. They do not want to spend a lot of time explaining the product to you, so it's up to you to learn about the product with little or no help. Sometimes, you can get information from other folks who know the product or are learning about it—for example, tech support personnel, trainers, and customer service reps. Save the limited time you'll get from the developers for the hard questions.

The Family Member Test

A side effect of your career as a technical writer is that your family and friends will assume that you're a computer expert. After all, if you spend your time writing about computers, surely you can explain computers to them.

You may not think of yourself as a computer expert, but you'll likely find that you do have enough expertise to help. If you're writing documentation for the general public, consider using the Family Member Test—whether the document you're writing will make sense to a family member who isn't a computer expert—to determine whether you're on the right track.

Writing ability

Writing skills are an essential component of technical communication. It seems terribly obvious to say that a technical *writer* should be able to write, but it's necessary to point this out. Enormous technical knowledge is not a substitute for writing ability; if you're not convinced, try reading material written by programmers sometime.[1]

The ability to break down complicated information into content that is appropriate for the document's audience is very important. Chapter 6, "Finally—it's time to start writing," covers this in detail.

The technical writer's mission is to create content (text and graphics) that communicates information to the reader. It's hard to identify exactly what makes good writing, but here are some general guidelines. Writing should be:

- Clear
- Easy to understand
- Not subject to misinterpretation
- Concise
- Easy to follow

It's easier to say what writing should *not* be:

- Confusing
- Redundant
- Wordy
- Poorly organized
- Inaccurate

Miss Thistlebottom was right...

An important component of clear, easy-to-understand writing is using correct grammar and spelling. As a technical writer, you must be able to produce content that's

1. *Some* programmers are excellent writers.

grammatically correct and doesn't have any spelling errors. Notice that we say "produce." Nobody writes error-free prose during a first draft. But you should spell-check, review, and proofread your drafts before anyone else—including your editor[1]—sees them.

However, following the standard rules of language comes with a significant caveat. In some cases, adhering to the most formal rules of writing is not the best method for communicating with readers. For example, assume that you are writing a guide that tells workers on an assembly line how to use time management software. Which of the following steps would be better suited for that manual?

> **Select the name of the building in which you work.**
> **Select the name of the building you work in.**

Even though the first example would probably make your high-school English teacher happier, it sounds pretentious. Your readers probably have well-tuned B.S. detectors, and it's unlikely that their priorities include avoiding a preposition at the end of a sentence. If creating a document with a casual tone will improve your reader's comprehension (and your credibility as the author), that goal is probably more important than following old-school rules of writing.

While writing, you will probably be required to follow style guidelines, either from a standard book (such as *The Chicago Manual of Style*) or from a company style guide. Style guidelines specify how you should capitalize section headings, for example, or punctuate bulleted lists.

---

1. Assuming that you're lucky enough to have an editor.

NOTE: You probably won't agree with all of the guidelines your company follows. Don't waste your time arguing about them, though, unless you can show that your preferred style improves reader comprehension enormously. Otherwise, you're just imposing your personal style preferences on the company. Arguing about nitty-gritty style issues is a great way to waste time and make yourself look unproductive.

If your writing skills need work, practice writing about any topic. Writing recipes, for example, tests whether you can remember to document every step in the process. Ask a friend to try your recipe and see whether you forgot any steps.

By far the best way to improve your writing skills, though, is to write a lot and have someone else critique your writing. Many fiction writers belong to writing groups; you might consider finding a few other new technical writers and establishing a similar group.

Organizational skills

Almost every job requires some organizational skills, but technical writing demands quite a bit. You obviously need the ability to organize information in your document. But to succeed as a technical writer, you also need project and time management skills.

Planning a schedule for a book means scheduling your writing time, plus time for many other activities—creating art, getting drafts reviewed, editing, indexing, and preparing for final output. Depending on where you work, you may be responsible for all of these activities, or you may need to coordinate with others, such as editors, to add your book to their schedules. For example, if your company has a technical editor, notify the editor about the project so that your book is scheduled for an edit. If your company's editing is limited to peer reviews (another writer editing your work), find out who might be available to look at your document.

To ensure that you can meet project deadlines, it's critical that the time and resources required for each activity are clearly spelled out at a project's start.

All of this probably sounds a bit intimidating. After all, you want to be a writer, not a project manager! In many companies, entry-level or junior writers aren't required to create schedules; a senior writer or documentation manager does the planning work.

If you are the only writer in a company, however, you may have to handle scheduling right from the beginning. It's unlikely that you'll get any help from your coworkers (they probably don't know a thing about documentation), so scheduling and time management will be particularly important skills in this environment.

Many writers are confident about their writing and technical ability but are intimidated by the project management and scheduling requirements of the job. Consider taking a seminar on project management and be sure to maintain a calendar or spreadsheet with scheduling information. Project management software can also be helpful, especially for large projects (for example, projects with lengthy schedules, multiple deliverables, or many people involved).

Strong detective (and people) skills

To create content, you need information. And unlike writing a novel, you can't just make up that information (although creative extrapolation is often required when you can't get any information—more on that in Chapter 6). Where do you get the information you need? Sources include technical specifications, prototypes, and product developers, as explained in Chapter 5, "Getting information."

Often, the problem is not getting information but identifying what information is relevant. You might have to ferret through piles of technical specifications to find exactly what you need, so filtering information is an important part of technical writing.

Sometimes, you need information from a product developer or a subject matter expert (SME, which is pronounced "smee") who is extremely busy and may not want to make time for you. Learning how to communicate and get what you need from busy technical experts is a skill you cultivate over time. (Chapter 5, "Getting information," contains some helpful tips on how to extract information from documents and from people.)

Now that you have an idea of the skills you need as a technical writer, let's take a look at what you do with those skills—the technical documentation process.

2 An overview of the technical writing process

There's more to creating documentation than just writing (designing graphics and editing, for example). As a writer, it's important to be aware of the other activities that are required—you may be asked to complete some or all of those tasks, particularly at a small company or in a small documentation department.

If you have a department that includes a technical illustrator, a technical editor, and a template designer, coordinate with them to ensure that your project gets on their lists of things to do.

Creating technical documentation, at a minimum, requires that you and your coworkers:

1   Identify the needed *deliverables*—the final products you turn over to the client at the end of the project. Deliverables can include books, online help, HTML, tutorials, wizards, and so on.

2   Write an outline (for a book) or a plan (a list of topics for online help) for each deliverable.

3 Create an overall project plan with a list of tasks for each deliverable and a draft schedule. Tasks include outlining, writing, editing, reviewing, taking screen shots, creating graphics, and so on.

4 Design the book's template or get the correct template from the template designer. A *template* is a file that contains all of the paragraph styles, page layouts, and other formatting elements for your document files. A template may also contain definitions for a document's structure; for example, the structure may require that every list have at least two list items.

5 Write the content.

6 Create the graphics or work with the technical illustrator to coordinate this task.

7 Edit the information and then have the material reviewed by a peer or technical editor. Make the changes noted.

8 Get the information reviewed by SMEs. Make the changes identified by the SMEs.

9 Index the document or give it to the indexer.

10 Produce the material (that is, clean up the formatting and get everything ready for the printer or the web or final delivery format).

The preceding process only scratches the surface. In real life, you'll get approximately halfway through step 6, at which point you'll discover that the developers have added a slew of new features to the product. You'll go back and document those features, test them against the software, and then discover that the developers also took out a couple of features without telling you. You check with a friendly developer around the corner and discover that those features are just "temporarily disabled"; development found some bugs but expects to correct the problems before the final release.

At this point, you have to make a decision. Do you assume that they will restore the features before the release, or do you delete the documentation? Or do you hedge your bets by making a copy of the information but removing it from the documentation for now? Because every project is full of happy surprises like these, this overview of the documentation process can only give you a general idea of how things work.

The rest of this book explains the steps in the documentation process in detail. So buckle your seat belt...

3 Very necessary evils—doc plans and outlines

What's in this chapter

❖ What's a doc plan?

❖ Outlining—it's not just for high school papers anymore

You've got the job, and you're ready to start a new documentation project. All you need is some information about the product, and you're ready to start writing— writing a documentation plan and an outline, that is.

You probably want to start writing the documentation instead of spending time on plans and outlines, but up-front planning and analysis are critical to good documentation. Formulating doc plans and outlines for your manuals is the best way to complete this analysis. The time you spend planning your documentation is an investment; you'll find that good planning saves you a lot of time later in the project. In other words, you can plan at the beginning, or pay for it when you flounder in the middle of the project.

NOTE: If you're using complex documentation development strategies, such as single sourcing or database publishing, planning is *not* optional.

To create a documentation plan and outline for each deliverable, you need some information about the product. Most often, you get access to a prototype of the product or a technical specification ("spec") that lists the product's features. Although early prototypes and specs *never* give you all the information you need (and frequently are inaccurate because the product has changed), they usually provide enough information to start the project. Chapter 5, "Getting information," explains how to round up information.

Clients aren't always where you expect them
A client is your customer. If you're a freelance writer or if your company is creating documentation for another company, the identity of the client is obvious. But as a staff writer in a software company, your client is probably inside the company. It may be your documentation manager, or it may be the vice president of development in a small company.

However, your ultimate customers are always the same—the readers of your documentation. Don't forget that your goal is to deliver information to them. |

What's a doc plan?

A documentation plan describes all the components of a documentation project. For a sample documentation plan, refer to Appendix D. A doc plan should contain information about the following:

- Product description—a brief summary of what the product does. This might only be a sentence or two: "The FoogleGarber software lets users ensure that their privacy is maintained as they surf the web. FoogleGarber rejects all cookies and other identifying characteristics requested by a web server."

- Audience—characteristics of the product's users; in some cases, there may be more than one type of user. You should provide as much information as you can about each type of user. This should include their education level, demographic information, level of technological expertise, and the features that each type of user will use. The more you know about your audience, the easier it will be to write documentation that meets their requirements. See "Audience, audience, audience" on page 76 for more information.

- Deliverables—the names of the documents you will create, a brief description of each, and how they will be delivered (printed manuals, online help, HTML, and so on).

- Receivables—what the writers and other documentation team members need from others (such as the latest version of the product, access to the developers to get questions answered, existing documentation, and a template from the template designer).

- Style—what style guidelines will be followed (such as in-house style guidelines and *The Chicago Manual of Style*).

- Tasks—list of the actions required to complete the project and who is responsible for them (for example, the documentation team will handle information gathering, writing, editing, production editing, and indexing. The product developers will provide technical assistance, review content, and ensure the accuracy of content).

- Tools—what tools (such as desktop publishing and graphics software) the documentation team will use to create the documentation. Chapter 4, "The Tech Writer's Toolbox," provides more information about tools.

- Schedules—a schedule for each deliverable (see "Any formulas for writing doc plans?" on page 40 for more information about estimating dates). A schedule should also include dates for editing, illustration, indexing, and preparation for final output (print, online, or both).

Sometimes, the doc plan itself is a deliverable, particularly if you're a freelance or independent writer. Even if you're not required to deliver the doc plan to your client or your internal customers (such as the development team), it's a good idea to let them see the doc plan to ensure that everyone understands what they need to do to keep the documentation deliverables on schedule.

Doc plans for external clients

If you are producing a documentation plan for an external client, the plan should also include the following items:

- Copyright—who will own the copyright of the completed documentation. It's critical to spell out whether and how the copyright is transferred from the writer to the client company. If you're in business on your own, get advice from an attorney on how to word this section of your proposal or documentation plan.

- Cost—a breakdown of the cost for each deliverable (see "Any formulas for writing doc plans?" on page 40 for more information). Cost may be a fixed price or an hourly rate.

- Disclaimer—All doc plans, contracts, and legal paperwork exchanged with a client should state that technical accuracy is the responsibility of the client or developer. For example, you could add the following text to your doc plan:

  "*Technical Communication Company* will make a reasonable effort to ensure the technical accuracy and completeness of the materials delivered to *Client*. However, *Client* is ultimately responsible for the content of the documentation. *Client* will review documentation for accuracy and completeness, and identify required changes to *Technical Communication Company* before the final delivery."

- Terms—payment schedule, cancellation policy, and other legal information.

Who writes the doc plan?

In a group of writers or a documentation department, a senior writer or manager usually writes the documentation plan.

However, if you're the only writer at a company, you're it. Take the time to write a doc plan—even if it's an informal document you don't show to anyone else. As you create the plan, you're forced to think about schedules, audience, and so on. Performing this analysis early in the project helps prevent nasty surprises later.

If your plan requires commitments from other people, be sure to consult them, especially before you publish your schedule. A double-booked editor is not a happy editor, and you don't want that (see Chapter 9, "Editors—resistance is futile").

Have your doc plan edited if you plan to turn it over to an external client—particularly when it's part of a proposal. A grammatical error or misspelling in a doc plan isn't going to improve your chances of getting the project, and it makes you look bad.

Any formulas for writing doc plans?

A doc plan contains lots of numbers (page counts, schedules, and so on), so you might wonder whether there are any formulas to help you calculate those numbers.

The Society for Technical Communication (STC) provides these estimates:

- 8 hours per completed page
- 8 hours per online help topic
- 40 hours per one hour of instruction time

For the STC's definition, a "completed" page includes all components of the writing process:

- Writing
- Editing
- Indexing
- Template design and document formatting
- Production

In your particular environment, you may find these numbers too high or too low.

To calculate page count, you might estimate that each task a user performs requires approximately two pages of documentation. Then, estimate how many tasks you need to document and throw in a few extra pages for the front matter and back matter. Creating these estimates is a bit of a black art, but this shouldn't keep you from creating estimates. At the end of every project, be sure to look back at the estimate. If the estimated numbers don't match the actual numbers, you need to identify why the estimate was inaccurate. As you complete more projects, you'll gradually develop a feel for how much time a project really takes and be able to improve the accuracy of your estimates.

Even experienced writers cannot pinpoint exact page counts or days of work with certainty. Instead, use your estimates as an indicator of what to expect. If you think that some factor, such as an extremely complicated product or a large number of changes, will increase the time required, be sure to account for that in your estimate. It's usually better to overestimate a little than to underestimate. The old cliché "underpromise and over-deliver" applies here.

If you have a drop-dead deliverable on a very aggressive schedule, putting together your plan will give you an idea of just exactly how much trouble you're in. Your plan can give you the ammunition you need to convince your manager that you need some temporary help.

Outlining—it's not just for high school papers anymore

You probably have bad memories of being forced to create outlines for term papers in high school. Well, those outline-writing exercises are about to pay off—writing an outline for a manual is a very important step in the technical documentation process. Outlining requires you to think about what needs to be covered in a manual, and that sort of early analysis is essential for good documentation.

In high school or college, you were probably quite capable of writing a 5- or 10-page paper without an outline and thought that your instructor was being difficult by requiring an outline. In reality, you did create an outline even if you weren't made to turn it in; the outline was in your head. But now we're talking about writing manuals that run 50, 100, or 500 pages. You will *not* be able to keep the outline for all of that in your head.

Creating an outline will help you break down the manual into manageable chunks. Let's face it: beginning a big documentation project is intimidating. If you create an outline, you can focus on just one section of the document at a time and pretend that the rest of it doesn't exist.

What goes into the outline?

To write the outline, you need to know what the users need to accomplish with the product. Your best bet is to play with the prototype, if one is available. You should also review the spec (if it exists) and read every bit of information you can get your hands on. This might include product design documents, marketing plans, business plans, and interface design documents. User documentation for related products is also useful for gathering ideas about what your document should contain. See Chapter 5, "Getting information," for details about information sources.

Based on your research and the information you find in the existing documents, figure out what the users do with the product. For example, if the product you're documenting is graphics software, users often convert graphics from one file type to another; if the product is a printer, users will want to know how to print on envelopes. This is where your ignorance (or ability to fake ignorance) of a product is really useful. As a new user, you need to figure out how to complete certain tasks. Eventually, you'll write down what you discover for other users. Use the mindset of a new user while you figure out what needs to be documented.

Once you have a list of tasks, you'll need to organize them into an outline—or outlines, if you think that the tasks would be better suited to more than one deliverable.

How many deliverables should there be?

Here are some things to consider while determining how many different deliverables there might be:

- *Do you have different audiences with different information requirements?* For example, it's common to deliver a book for regular users and another one for system administrators so that you can separate out the day-to-day tasks from the high-level tasks that only users with special rights on the system can perform.

- *How much information do you have?* If you're looking at delivering about 100 pages of information, you can probably put everything in a single book even if you do have multiple audiences. However, if you have a total of 1,000 pages of documentation for a single product, you may want to separate the material in several books just so that your readers won't strain their arms trying to lift your tome. Dividing up the information into several books makes it easier for them to find what they're looking for.

- *Do you need to deliver information in different media?* One common approach is to deliver interface help ("The Add button does this.") or task-based information in the online help but to provide conceptual details in the printed documentation.

- *Are there delivery issues that you should consider?* For example, even if most of the documentation is delivered on a CD, you should always provide printed installation information. Otherwise, end users are caught in a Catch-22—if the installer doesn't work, they can't check the documentation because the documentation hasn't been installed yet!

Writing the outline

Some word processing programs (Microsoft Word, for example) have outline functions that can handle most of the formatting for you.

Instead of using a rigid outlining structure with Roman numerals, consider using headings, bullets, and sub-bullets, as Figure 1 on page 46 shows.

There's no single way to develop the outline. Do whatever makes sense to you.

Your outlines can be part of your documentation plan or stand-alone documents, but no matter what, be sure to get the outlines approved. Your manager or a more experienced writer can help identify any items you may have forgotten. If you're freelancing, be sure to get your client's approval of your outlines; this will ensure that your client understands where you're going with the documentation early in the process.

Chapter 1: Entering job data

This chapter will contain the following sections, which will include screen shots of the graphical user interface (GUI) windows a user encounters while completing procedures:

- Importing data—procedure explaining how to import data from a spreadsheet or from a tab-delimited file exported from legacy software.

- Manually defining a job—provides a flow chart showing all the characteristics a user can define manually if a spreadsheet or tab-delimited file is unavailable. The chart will include cross-references to detailed procedures, which are as follows:

  - Defining process types—procedure explains how to define the departments and areas through which a job flows. Section will also explain how to change and delete process types.

  - Defining job sites—procedure explains how to define the job sites with which work may be associated (for example, Plant A in North Carolina). Section will also explain how to change and delete job sites.

  - Defining employees—procedure explains how to define employees that work at each job site. Section will also explain how to group employees according to job functions and how to change and delete employee information.

Figure 1: *Outlines don't have to be formal*

4   The Tech Writer's Toolbox

What's in this chapter

- ❖ Content/text development tools
- ❖ Graphics software and clip art packages
- ❖ Help or web authoring tools
- ❖ File conversion utilities
- ❖ Other helpful software
- ❖ Computers and ergonomics

Now that you have a doc plan and an outline, there is one last thing you need to consider before you dive into writing manuals: do you have all the tools you need to complete the job?

This chapter explains the software you need in your Tech Writer's Toolbox.

NOTE: See Appendix C, "Tools information," for a list of programs and the companies that make them.

Content/text development tools

The most important tool in your toolbox is the one you use to write the text. There are many word processing, document processing, and desktop publishing packages out there, and each one has its advantages and disadvantages.

If you're working in a corporate environment, you probably won't get to choose this tool. You'll use whatever your manager hands you—and like it (maybe).

Many technical writers use a basic word processing program such as Microsoft Word. Although Word is adequate for short business documents, it is not designed for long, complex documents, and it often becomes unstable when you try to maintain very lengthy documents.

For basic technical writing, you need a tool that can:

- Handle many embedded graphics
- Number steps and figure captions automatically
- Create complex tables
- Maintain cross-references
- Generate indexes and tables of contents

Based on this list, a good choice is Adobe FrameMaker software. FrameMaker is specifically for creating and maintaining long technical documents. Taking advantage of FrameMaker's powerful document processing features can greatly improve your productivity by reducing the amount of time you spend on tedious document maintenance. Other tools in this category include Ventura.

If you plan to create different kinds of output from your document (for example, online help and HTML), keep in mind that you'll need to find an output path from your source document format to the other formats. See "File conversion utilities" on page 53 for details.

If you're producing graphics-intensive shorter pieces (such as a newsletter), you might consider one of the desktop publishing packages, such as PageMaker, QuarkXPress, or InDesign. These packages are less oriented toward book production but are better than FrameMaker at producing full-color, highly designed documents. If your documentation needs to look like a very expensive annual report, these tools may be right for you.

Instead of using a document processing or desktop publishing application, you may use a tool for authoring in Extensible Markup Language (XML). XML is a specification for storing structured content as text. *Structured content* means that the text in your document must follow a specified flow. For example, structure definitions may specify that a chapter must begin with a title followed by one or more paragraphs and that all lists must contain at least two items. See Chapter 14, "Structured authoring with XML—the next big thing," for more information. XML development tools include Corel XMetaL and Epic Editor by Arbortext.

Graphics software and clip art packages

The illustrations, screen shots, and other graphical elements you use in your documentation need to come from somewhere. You need several kinds of graphics software:

- Drawing—to draw shapes (particularly for flowcharts and technical illustrations).

- Screen capturing—to take snapshots of items on your computer screen. If you're writing about software, you'll need pictures of the software, and screen capture software does that quite well. Some operating systems (such as Windows) have basic screen-shot capability already built in. See "Displaying information from your computer screen" on page 114 for information about taking screen shots through the Windows and Macintosh operating systems.

- Graphics processing/editing—to change a file's format, to touch up and crop graphics, and so on.

Contemplate that template

Regardless of what content development tool you use, you'll probably write content based on a template. Most documentation departments have specialists who design how content looks: the fonts used, the spacing between paragraphs, and so on. All that design information is stored in a template file. Before you begin writing, you create a blank document and import the formats from the template into your new file. The template file itself may document how to use the formats, or there may be a separate guide with that information. Also, there may be multiple templates in your department (one for user guides, one for developer guides, and so on).

Some templates do more than define formatting—they also define the structure of a document. Structure definitions determine what content is valid at a particular point in a document. Chapter 14 describes working with structure in detail.

As you write content, it's not advisable to create new formats; in fact, that can get you into a lot of trouble with your manager, the template designer, and the production editor (as explained later in Chapter 11). Instead, explain to the template designer why you need a format that is not in the template. The designer will evaluate the request and may add a new format to the template; that way, the new format is available to all writers. However, be prepared for the designer to tell you what format you should be using instead of the new one you want. If template designers added a format every time a writer requested it, the template would probably have hundreds of formats and be impossible to use.

Templates are also common with other applications technical writers use: graphics programs, help authoring tools, and file conversion utilities. Regardless of what kind of program it's for, a template generally has the same purpose—to ensure consistency in presentation, which is an essential component of good technical communication.

Some software packages (such as Paint Shop Pro) can handle both screen captures and graphics processing. However, most drawing tools don't provide graphics processing and vice versa.

Paint Shop Pro is popular for taking screen shots on Windows machines. For the Mac, Snapz Pro is available. For UNIX, try xv.

If you need to create flowcharts, consider Microsoft Visio.

For drawing line art, you'll need a package such as Adobe Illustrator or MacroMedia FreeHand.

Finally, for graphics processing or editing, you need something like Adobe Photoshop or Paint Shop Pro.

Some graphics programs come with clip art packages, which can be helpful when you need icons in your text (such as a stop sign to flag warnings). If you don't have much drawing ability or access to a graphic designer's services, a well-chosen clip art package can prevent quite a few headaches. Keep in mind that you want a high-quality selection of clip art and that some clip art packages have restrictions on the number of times you can use an image and where (for example, some permit you to use the images on the web only).

If you buy a professional drawing program (such as Illustrator), it probably won't include clip art, but you can buy clip art separately. Corel offers several clip art packages.

Help or web authoring tools

If you're developing online help or web-based materials, you'll need a tool to write the help or HTML. For online help, you can use tools such as RoboHELP, and to write HTML, you can use tools such as Dreamweaver or GoLive.

If you're going to create documents for more than one type of output (for example, hard copy and HTML), your best bet is to use a text development tool that has conversion capabilities (or that's compatible with a third-party conversion tool, as explained in the next section).

File conversion utilities

There are many third-party tools that convert word-processing files to other formats, including online help, HTML, HTML Help, JavaHelp, and XML. Converting the material intended for print means that you don't have to spend time re-creating a different type of output in another authoring tool. Conversion tools include Quadralay's WebWorks Publisher and Omni Systems' MIF2GO.

Another online file type is Adobe's Portable Document Format (PDF). A PDF file maintains the formatting of a hardcopy document, but it gives you the benefits of online documentation, such as hypertext cross-references. To convert documents to PDF format, you'll need Adobe's Acrobat Distiller. The Acrobat Reader, which lets users view PDF files, is available for free from Adobe.

The ability to create multiple types of output from one set of files is called single sourcing. See Chapter 13, "Single sourcing," for more information.

This software is hard to use! Help!

Some of the tools listed in this chapter—particularly for text development, graphics, and file conversion—are not applications you can learn quickly (or even easily). Getting formal training is the best option, but it's expensive. Many software manufacturers' web sites list companies that train people how to use their software. Your local chapter of the Society for Technical Communication (STC) may list (and even sponsor) training in your area. Also, check with your local community college about software-related continuing education classes. If you're lucky, your employer may reimburse you for the cost of training.

If training is not a possibility for you, consider getting a third-party reference and a workbook with exercises. For example, Scriptorium Press (www.scriptorium.com/books) offers the FrameMaker 7 Workbook Series.

Other helpful software

There are some other software packages you'll need as a technical writer. These tools, which aren't limited to the technical writing profession, include the following:

- Compression utility—Files can become quite large, particularly if they include graphics, so it's a good idea to have a tool such as WinZip (PC) or StuffIt (Mac) that compresses files before you send them to a coworker or the client. Compression is essential when sending large files via email.

- Communication software—You'll need a tool for email (such as Microsoft Outlook or Eudora Pro) and file transfer protocol (FTP). Sometimes, files are too big to send through email as attachments. FTP software lets you transfer the files over the Internet without using email. Some web browsers have built-in FTP capabilities, so you may not need a separate program. A good rule of thumb is not to send an email attachment larger than two megabytes (MB). Don't forget to use your compression utility for larger files sent by FTP, too.

  On the PC, you can use commands at an MS-DOS prompt to send files via FTP, but many people prefer using software with a user interface (such as WS_FTP). For the Mac, try Fetch.

- Project management/time-tracking tool—Having software that tracks your schedule can be helpful, particularly when several people are working on a project. Even if you are the only person on a project, laying out your schedule is useful. Another option is using a web-based calendar that all team members access. That way, everyone can see when deliverables are due. You can create such a calendar for free at web sites such as Yahoo! (http://www.yahoo.com), which offers calendars through its Yahoo! Groups service.

- Encryption software—If you're sending confidential information over the Internet, consider using encryption software such as Pretty Good Privacy (PGP) for your email. Keep in mind that both the sender and the recipient will need the software.

NOTE: In your company, you may be required to use specific tools for email, project management, and so on. Your company may also have specific configurations for the tools.

Computers and ergonomics

To run all the software just mentioned, you need a computer with a powerful processor and a lot of memory (random access memory, or RAM). Most newer computers have more than enough processing capability and memory to run the programs you'll use (and that includes any software that you may be documenting). If you're using an older computer, though, it's a good idea to check the system requirements of applications to ensure your computer meets the specifications.

You're going to spend a lot of time looking at text on your monitor, so get the biggest one you can afford (or talk your boss into buying a big one). At an absolute minimum, your monitor should be 17 inches. This will help minimize scrolling through your pages.

Also – Double Monitors

Because technical writers do spend so much time in front of the computer, it's important to have an ergonomic work space to prevent repetitive motion injuries (including carpal tunnel syndrome) and other problems (such as eyestrain).

Some things to consider about your work environment and how you work in it include the following:

- Height and position of your chair and keyboard
- Posture
- Lighting

For resources about ergonomics, see "Ergonomics" on page 256.

5 Getting information

What's in this chapter

- ❖ Technical specifications and other development documentation
- ❖ Prototypes and software under development
- ❖ Legacy documentation
- ❖ Developers and subject matter experts
- ❖ Interviews with users

Your primary task is to give people the information they need to use technology. Ferreting out that information can be the most difficult aspect of technical writing. For this reason, technical writers are sometimes called "information developers"—they develop useful information from various obscure sources.

This chapter explains several sources from which you can extract information.

NOTE: In general, you need some information about the product to write a doc plan and outlines for manuals, so you may have access to at least one source very early in the project. As work progresses on the project, you will probably use most of these sources.

Technical specifications and other development documentation

A technical or functional specification—known as a *spec*—is a document written by a product's developers. It explains the product's purpose and how it works. Typical information in a spec includes:

- Lists of menus and menu choices (software)
- Mock-ups of the interface (software)
- Illustrations of components (hardware)
- Lists of features or proposed features
- Information about how the product processes data (for example, how a connection to a database works)
- Information about the product's components
- Changes from the previous release
- Schedules for the product's release

The benefits of a spec

A good spec can provide you with an excellent overview of the product and answer many basic questions about the product. When the spec answers your questions, you don't have to track down a developer or SME to get your information, and that makes everybody happier.

The drawbacks of a spec

Most specs don't contain all the information you need, are inaccurate, and aren't updated to reflect a product's ever-changing functionality. Often, the spec doesn't exist at all.

In short, a *really* good spec is a mythical creature. If you ever see one, please contact your nearest tabloid—you could probably make a lot of money selling the story about your encounter.

> ## Organizing menu information: the spec does not always know best
>
> Because a software spec often lists every menu and menu choice in the order they occur on the interface, you may be tempted to write your document in a similar fashion—but don't! Instead, focus on the tasks a user performs and the order in which the user performs them. This may or may not match how the interface is organized.
>
> For example, many software programs have a **Print** menu choice on the **File** menu. If you wrote the manual according to the **Print** choice's order on the interface, information about printing would come early in your book because it's on the first menu, **File.** In reality, however, printing may be one of the less important tasks a user performs, so it may not need such prominence in your manual.

Prototypes and software under development

The term *prototype* is used a little differently in hardware and software. A hardware prototype often does not have all the working parts; it may consist of the product's "shell" without all the internal components. A software prototype is usually a "proof of concept"; the software performs more or less the functions that the final software will perform, but the interface is probably not what will appear in the final product.

As a writer, you will find hardware prototypes very useful because you can often see how the product will look and can infer what will happen "under the covers" from the design. But software prototypes can be a problem because the documentation must be very specific about how to manipulate the software's interface. In fact, most users (if they've ever thought about it) equate the interface with the product. The software developer will tell you, "Oh, the functionality is the same; we just tweaked the GUI."[1] But your end users don't care about the functionality; they care about how things look. So, a "tweak" in the interface can result in catastrophic rewriting requirements for the documentation.

Slightly better than prototype software is prerelease software. Usually, the software goes through a few stages, as shown in Table 1:

Table 1: *Software release cycle*

Development stage	Status
Alpha	Software works, sort of. Some major features are missing or not working.
	Expect alpha software to crash your system at regular intervals and possibly corrupt it. If at all possible, keep alpha software and your documentation files on separate computers.
Beta	Software works, mostly. All major features are sort of working. This version is often made available to customers for testing.
	Expect beta software to work with minor glitches. It may crash occasionally, but you should be able to identify what causes the crashes.

---

1. GUI (pronounced "gooey")—graphical user interface. Otherwise known as the windows that the user sees on the screen.

Table 1: Software release cycle (continued)

Development stage	Status
Release Candidate (RC)	Software is done—for the most part. All features are working and major bugs have been cleaned up. It's made available to more customers for testing. RC software should be stable with no crashes.
General Availability (GA)	Software is done and shipped to customers.

Many companies have additional components in their release cycle. Products may go through several alpha versions (Alpha 1, Alpha 2, Alpha 3) before moving on to beta status. Often, Release Candidate status requires a freeze on interface changes and no code changes except to correct bugs. In other words, RC status means that no more features will be added.

Of course, these release cycles are only guidelines, and they are for the most part breached more than they are followed. For example, it's common for companies to release software that's not quite ready to hit a particular shipping deadline. (Most often, this release coincides with the end of a quarter and has various implications for what's known as "booking revenue.")

The benefits of prototypes and prerelease software

Early "drafts" of the product in the form of prototypes and alpha or beta software are an essential source of information. If you're documenting software, you must have a copy of the program to write and test your procedures (and to take screen shots for your document). The same applies to hardware prototypes.

The drawbacks of prototypes and prerelease software

Like a spec or any other source of information, if the prototype or software you're using is not the latest version, the information you're writing could very well be inaccurate. Be sure that when the development team makes changes, you get the new version or are at a minimum notified of the changes.

NOTE: A good way of keeping track of changes is to get access to the developer's bug tracking system, which can provide invaluable information about what's being changed. At some companies, writers can add bugs they find to the tracking system.

If you are taking screen shots of a software interface or are drawing hardware diagrams, it is especially important that you wait a bit for the product to stabilize and then take the screen shots or draw the illustrations. Keeping in close contact with the product developers can help you figure out when it's safe (well, safer) to create graphics. Inevitably, you will have to retake screen shots or update drawings due to last-minute changes. But you can minimize the amount of rework you'll have to do by waiting as long as possible before creating illustrations or taking screen shots. Instead, insert placeholders that explain what a drawing or screen shot will show.

Small software companies are especially notorious for changing things up to the very last minute. If you're working at one of these companies, try to educate your managers about why these changes will cause problems for your documentation efforts.

Of course, a changing product can also cause difficulties for your text. One good approach is to ask the developers which parts of the product are more stable so that you can write about those first. Then, as other parts are completed, you can document them. This can save you from having to rewrite a section several times as the product changes.

TOP SECRET

Many products are considered confidential while they're under development. Companies don't want their competitors to hear about the new product or its features.

If you are working on a product whose features (or existence) have not yet been announced, you can expect some security requirements. Generally, you should avoid discussing the specifics of what you're working on with anyone who is not involved in the project.

If you are a freelancer, you'll probably be asked to sign a nondisclosure agreement (NDA). The NDA spells out your obligations to keep the information you'll get confidential and to safeguard your client's proprietary information. Most NDAs are straightforward, but read each one. In some cases, clients will put intellectual property information into their NDA. The intellectual property and copyright agreements should be separate from the NDA, which should focus only on how you must handle confidential information that the client divulges to you.

If you're working on classified government information, expect much more stringent requirements. But nothing could top the company that required its technical writers to hide a prototype printer when they weren't working on the documentation. The writers were required to cover the prototype with a large box whenever they left the room so that no one could look in and see the printer. In addition, the development had to be done in a room with no windows, and the door had to be locked when the writers were not in the room. This is a true story. Really!

Legacy documentation

If you're documenting a new version of an existing product, there's a good chance that documentation already exists for the previous version. In some cases, you can use the existing content by updating the text to reflect changes and added functions.

The benefits of legacy documentation

Legacy documentation can help when you're drawing up doc plans for new documentation—you can ask your clients and the documentation's readers what they like (and don't like!) in the existing docs and then use that feedback to improve the documentation for the next version. If the documentation is well written, you can use it as a foundation for the new material and add, update, or delete information as necessary. Also, working with what's already been written can save a lot of time.

Reviewing documentation for similar products can also be helpful. For example, if you're documenting a printer that's in the same product line as one that's already on the market, you may be able to crib some of the text from that product for the manual you're developing.

The drawbacks of legacy documentation

Legacy documentation can be a problem when it's badly written. Reworking existing bad content into something useful takes as long as writing quality content from scratch. If the product has changed significantly since the last release, the legacy documentation may not have enough relevant information to make reuse worthwhile.

You'll run into another problem with legacy docs if you're new in a job or when you pick up a new client. You may review the legacy documentation and discover that it's poorly written, disorganized, and not useful for readers—but your client or manager thinks that the existing manuals are great and just wants you to "make a few updates" for the next release.

You'll need your diplomatic skills to solve this type of problem. You could try rewriting a brief section and explain the improvements you've made.

The best way to avoid this situation, though, is to ensure that you find out *before* you take the job what legacy documentation exists and what the client or manager's opinion of those documents is. Review the material before you accept the job and make sure that you and your potential employer agree on what needs to be done.

Developers and subject matter experts

The people who are developing a product—or the SMEs very familiar with a product—are the most important source of information a writer has. Not only do developers create the prototypes, but they also know which features work, how features should work when they don't, and what changes are on the horizon. Good communication between developers and technical writers is essential to the success of a documentation project.

The benefits of developers and SMEs

Developers who promptly answer your questions and review content on schedule are your best allies on a documentation project. Their review comments ensure that your documentation is technically accurate, and their knowledge of how the product is shaping up can cut the amount of time you spend rewriting material. If the developers warn you that particular features are going to change, you can hold off writing the material about those features until the developers tell you the features are stable.

The drawbacks of developers and SMEs

Sometimes developers are so busy working on the product, they feel they don't have the time to review document drafts. (In some extreme cases, developers may not place much value on documentation, so they don't bother to review it at all.) Reviews then run late, or they are very cursory and offer no real feedback. Sometimes, review comments focus on issues such as comma usage or capitalization instead of on the technical content.

To avoid problems with grammatical nitpicking, you should do three things:

- Ensure that the documentation drafts you deliver are grammatically correct and spell-checked and that they do not contain writing errors that will distract the reviewers.

- Remind the reviewers that you need their input on technical issues, not on grammar. If appropriate, tell them that an editor will review the material to ensure that it conforms to the company style guide.

- Create a sign-off sheet and attach it to the draft (or send it as a separate attachment to your email if you're sending drafts electronically). The sign-off sheet should specifically request that review comments focus on technical accuracy and completeness. Requiring reviewers to sign off on their reviews makes them more accountable (and can also provide you with some protection if technical errors aren't caught).

Getting a solid review from the developers is essential because only the developers know whether content is technically accurate. Without a thorough review by the development team or SMEs, your manuals will not be as useful to the product's users.

What are some of the ways you can ensure that developers give you the information you need? See the following sidebar, "(Almost) 30 ways to get information from developers."

(Almost) 30 ways to get information from developers

While many of the following suggestions are hardly serious, the basic message should be clear: establish a clear line of communication, be persistent but respectful, and remember that the developer is a person with interests and responsibilities outside of work.

1. Bring bagels and cream cheese to a morning review session.

2. Deliver documentation that requires only minimal changes from the developers. If the developers respect your work, they'll be more likely to deliver the information.

3. Figure out the developer's preferred method of communication and use it. If the developer prefers to talk at the water cooler, do it. If the developer prefers to be contacted by email, use email, even if his cube is across from yours.

4. Be respectful of the developer's time and other commitments. Try to group your questions instead of interrupting her constantly.

5. Consider using medieval torture devices, including the rack and the iron maiden.

6. Develop enormous expertise in your product. The product engineers are more likely to take your questions seriously if they think you know your stuff.

7. Restrict the questions you ask the developer to the really obscure stuff.

8. Brownies. And lots of them.

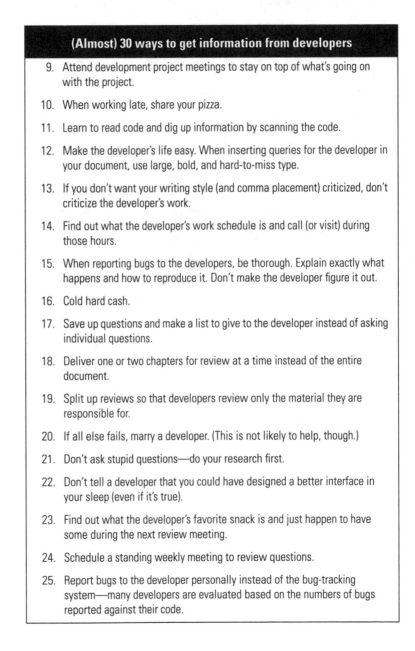

(Almost) 30 ways to get information from developers

9. Attend development project meetings to stay on top of what's going on with the project.

10. When working late, share your pizza.

11. Learn to read code and dig up information by scanning the code.

12. Make the developer's life easy. When inserting queries for the developer in your document, use large, bold, and hard-to-miss type.

13. If you don't want your writing style (and comma placement) criticized, don't criticize the developer's work.

14. Find out what the developer's work schedule is and call (or visit) during those hours.

15. When reporting bugs to the developers, be thorough. Explain exactly what happens and how to reproduce it. Don't make the developer figure it out.

16. Cold hard cash.

17. Save up questions and make a list to give to the developer instead of asking individual questions.

18. Deliver one or two chapters for review at a time instead of the entire document.

19. Split up reviews so that developers review only the material they are responsible for.

20. If all else fails, marry a developer. (This is not likely to help, though.)

21. Don't ask stupid questions—do your research first.

22. Don't tell a developer that you could have designed a better interface in your sleep (even if it's true).

23. Find out what the developer's favorite snack is and just happen to have some during the next review meeting.

24. Schedule a standing weekly meeting to review questions.

25. Report bugs to the developer personally instead of the bug-tracking system—many developers are evaluated based on the numbers of bugs reported against their code.

(Almost) 30 ways to get information from developers

26. Send your manager a note explaining the drop-dead date when you need review comments.

27. Copy your manager on correspondence to developers.

28. Copy the developer's manager on requests for information.

29. Have your manager talk to the developer's manager (a last-resort measure).

Interviews with users

You write a book for a product's users, so it makes sense that they could provide you valuable information during the writing process. Unfortunately, this is by far the most difficult information to get.

The benefits of interviews with users

By talking to users of a product, you can get firsthand information about how the product is really used. Users can tell you what they stumbled on (so you can be sure your document clearly explains the issue), and they can show you the most frequently performed tasks (which can help you figure out in what order tasks should be described in the documentation). Sometimes, an interview with a user will prove that the document's outline does not reflect the tool's use in the real world, so you may need to revamp the structure of the outline (and maybe even your document, if you've already done a lot of writing) to reflect reality.

The drawbacks of interviews with users

Tight budgets and aggressive schedules on documentation projects often make user interviews impossible— interviews are expensive and time consuming, whether the interviews are done in person, with paper or online questionnaires, or over the phone. Even if you manage to get access to customers for the interviews, you sometimes end up with individuals who aren't the product's true users. You may end up talking to the person who supervises the employees using a product instead of the employees themselves. Feedback from such an individual can do more harm than good because the supervisor's opinion on how the product is used (or should be used) can be vastly different from the employees' day-to-day reality.

For new products, you may not know who the users will be, which makes it difficult to interview them.

Although your mission is generally to act as a user advocate and to provide the information that the users want and need, there are cases where this might not give you the true picture. If you're documenting company policies and procedures in addition to how to use the product, you may need to talk with managers and supervisors to ensure that you get the appropriate information about corporate policy for your documents. This corporate information may conflict with how users would like to use the product.

6 Finally—it's time to start writing

What's in this chapter

❖ Audience, audience, audience

❖ Style and terminology

❖ Different types of content

❖ Dealing with the inevitable schedule changes

❖ Experience is the best teacher

Considering that "writer" is part of a technical writer's title, it's surprising how long it takes to get around to actually *writing* the documentation. Once you've created a documentation plan, drawn up outlines, and identified your information sources, you're ready to get started.

So, how do you get started? Basically, you need to perform the same tasks the product's user does, and while doing so, write down exactly what you do and what happens when you do it.

However, the documentation process is not as simple as just writing down actions and results. You need to provide background information, such as a high-level explanation of what the product does and why, and information about why the user might want to perform a particular action. It's usually much more difficult to create that information than to write down tasks because the background or conceptual information requires that you truly understand the product and its uses.

While writing your material, you need to tailor the content to the audience, break down a task into discrete steps, and decide when to use graphics and tables. This analysis is an important part of writing good documentation.

This chapter explains the many points you should keep in mind while writing technical documentation.

Audience, audience, audience

The real estate agent's mantra is "Location, location, location"; your mantra should be "Audience, audience, audience." To create a useful manual, you must address your audience at the right level. For example, if you're documenting a product used by workers who have little prior computer experience, you may need to explain basic operating system procedures—for example, the difference between a single and double mouse click. However, if you're documenting a tool used by C++ programmers, an explanation about single- vs. double-click would not only be inappropriate, but it would also annoy experienced computer users. When a document underestimates (or overestimates) its audience, users become frustrated with the manual and begin to doubt its accuracy and even its usefulness. Such a document is a failure.

Although it seems obvious that instructions on how to use the mouse aren't needed in programmer-level documentation, lack of attention to the target audience is a major problem in technical documents.

You've probably experienced this firsthand. Have you read a poorly written manual for a household appliance or a video cassette recorder? Imported electronics are especially notorious for bad documentation (with the added problems of poor translation from the source language into English).

There's a very good chance the guide was a failure because it was unclear and because it overshot its audience's knowledge or experience level. Manuals should help people use products—not make them feel stupid. Writing to the correct audience will prevent a great deal of user frustration.

K.I.S.S.

The K.I.S.S. principle (Keep It Simple, Stupid) definitely applies to technical writing, but keep in mind that writing for your audience does not mean that you should be patronizing or condescending. Your readers will see right through this. You should also be careful about making assumptions about your audience based on demographics or educational level.

The rule of thumb for technical writing is that you should write at an eighth-grade level. You can use software to check your documents for complexity (some software will even evaluate the grade level for you), but here are some general guidelines on what "writing at an eighth-grade level" really means:

- Use clear declarative sentences.

- Avoid jargon, slang, and idioms.

- If you need to use complex terms or acronyms, explain them.

- Use headings, paragraphs, bullets, and steps to structure your writing into manageable chunks.

NOTE: In Chapter 12, "Avoiding international irritation," you'll find out that these and other guidelines in this chapter also make it easier to translate your documents.

Inclusive language

Using inclusive language is sometimes derided as "political correctness." It's true that some phrases can sound awkward ("his or her"), but it's possible to write a document that uses inclusive language without being obvious about it. For example, consider these choices:

1. The WinkleWart product helps the user manage his finances.
2. The WinkleWart product helps the user manage her finances.
3. The WinkleWart product helps the user manage his or her finances.
4. The WinkleWart product helps users manage their finances.
5. The WinkleWart product helps you manage your finances.

The first two options are not inclusive. The use of the so-called "generic he" in the first sentence could be construed as including women. However, it annoys many readers, so it's best to avoid it—even if it's technically grammatically defensible.

Option 2 might be acceptable if you're writing for an all-female audience. Option 3 is inclusive, but it's glaringly obvious you're trying to be inclusive. Consider the fourth and fifth options. Both of these options let you avoid the "he or she" construction gracefully.

In a few cases, it will be difficult to write around using "he" and "she." In such cases, you can alternate between the masculine and feminine pronouns.

NOTE: Common sense should prevail. If you're writing a document about pregnant women, "he or she" is probably not appropriate.

Occasionally, you'll be asked to document a product that is targeted toward a particular group—perhaps women or a particular ethnic or racial group. Take your audience into account, but don't insult your readers by assuming that sprinkling a few ethnic names throughout the document will, by itself, make your document appropriate for the audience.

Analyzing your audience without spending a fortune

You'll find entire books available on audience analysis; see "Audience and task analysis" on page 250 for resources. But in the real world, you rarely have the time (or money) to send out detailed user questionnaires, interview prospective users, or use other sophisticated techniques. There are, however, a number of questions you can ask to get a sense of your audience:

- *What is your audience's educational background?* If you're writing for college professors, you can make some assumptions about their level of literacy and education.

- *What is the demographic profile of your audience?* For example, where do they live, how old are they, and is English their primary language?

If you're writing for teenagers, using the Temptations as an example is guaranteed to result in a document that's labeled "uncool."[1] If your audience has limited English proficiency, it's a good idea to use graphics to convey your main points. (You might also consider translating the documentation.)

- *What is your audience's level of computer knowledge (or knowledge about the hardware)?* Professional computer users, such as programmers or system administrators, have different documentation requirements than computer novices.

- *How much do readers know about the subject matter?* If you're documenting how to use a legal database, the lawyers who use it are probably experts on the content in the database. They are, however, unlikely to be computer experts. In this case, you don't need to worry about explaining legal terms, but you do need to explain computer terms. If your product is a database of legal information for the general public, you cannot assume that the reader will have legal knowledge or computer knowledge.

- *How motivated is your reader?* Does the user want to use the product? If you are writing about a digital camera, you can probably assume that most of your readers are interested in using the camera and want to learn about it so that they can take good pictures. But what if you're writing about a database application that's replacing a paper filing system in an office? The clerks who have been using the paper

---

1. Note to younger people reading this document: The Temptations were a very famous Motown group in the 1960s.

filing system for years know exactly how to use the old system. But now, they must learn a new, computer-based system, which may not be popular. In this case, your audience's motivation could be very low. The lower your audience's motivation, the easier (and shorter) your documentation should be. A more highly motivated reader is more willing to invest some time reading documentation to learn about the product.

- *What are the user's requirements?* What do the users need to do with the product? Do they need to know how to do everything or just a how to use a few important features? This question will help you assess whether your information should be strictly task-oriented (how to accomplish task X) or should include reference information (an exhaustive list of features, usually organized alphabetically or by some other scheme that doesn't address the order in which tasks are performed).

- *Do your readers have any physical characteristics or limitations that affect their reading ability?* These limitations will affect how you design and deliver the documentation. For example, older readers or readers with limited vision will not appreciate small, cramped fonts. Readers who are color blind will not notice that you've cleverly color-coded headings for them. If your audience includes people who are blind, you need to make sure that your documentation is comprehensible to someone who *hears* it or that it works in Braille.

Answering these questions will help you understand your audience and write documentation that better meets their requirements.

Style and terminology

Before you start writing, be sure to consult the style guidelines for your project. Most projects have editors that establish a project's guidelines—when to use bold type, when to capitalize words, what terms to use when referencing parts of the product, and so on. If you learn the guidelines before you start writing, you can prevent (or at least minimize) the misery of fixing repeated stylistic errors in your text.

Chapter 9, "Editors—resistance is futile," contains more information about style guidelines and working with editors.

Different types of content

You can divide up the content you need to create into several different categories. Again, you can find volumes of academic research on this topic, but here are some basic categories:

- Interface information
- Reference information
- Conceptual information
- Procedural information

The next few subsections describe each of these types of information and how you might handle them as you assemble documentation.

Interface information

Interface information explains the function of a particular part on a product. For hardware, you'll often provide interface information as an illustration or a photograph. For example, you show a larger view of the product, and then zoom in on a particular part (as shown in Figure 2). You can further point out specific parts with *callouts,* which consist of lines and labels ("Hula man" in Figure 2).

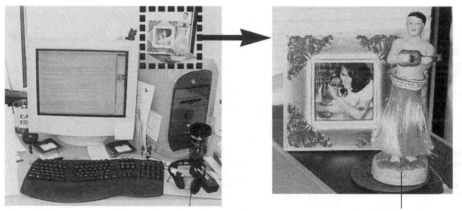

Hula man

Figure 2: *You can provide an overall view and then zoom in on the most important part of the picture in a separate image*

For software, you can provide similar labels electronically. Many applications include ToolTips, which are pop-up labels displayed when the user rests the cursor over a particular item for a few seconds (Figure 3).

Figure 3: *ToolTip in Eudora Pro email software*

If you're developing documentation for a software application, you can also provide context-sensitive help. The user can display a small window with a sentence or two explaining a particular window or button. As shown in Figure 4, context-sensitive help lets you provide more detail than ToolTips, which are normally limited to just a word or two.

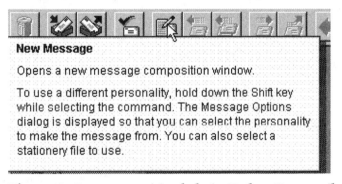

Figure 4: *Context-sensitive help in Eudora Pro email software*

Some writers provide a list of all the interface items in an appendix in the documentation. Consider whether the effort in assembling this information is worthwhile. It's not likely that your readers will read an appendix just to see all the interface items explained. However, linking online interface descriptions to the items on the interface gives immediate assistance to users.

Building context-sensitive help can be technically challenging. It requires that you work closely with the software developers and use some complex tools. But most help authoring tools have context-sensitive help capabilities built in, so you don't have to build it by hand.

Reference information

Reference information is data that readers need to look up (or "reference"). A dictionary is a great example of reference information. A standard dictionary provides an enormous amount of information about words, but it doesn't tell you anything about how to write sentences. So, reference information assumes that the reader knows what to do with the information; it provides content but no instructions. In technical writing, reference information includes glossaries and lists of functions for a programming language.

Writing reference information is not particularly difficult (although actually acquiring the information can be). Because the information is usually highly structured and organized in a way that's not up for debate (for example, alphabetically), you can create a skeleton for each reference item and then just fill in the required information.

The advantages you get from online formats, such as clickable indexes, full-text search, and hypertext links, are particularly useful when applied to reference information. In a book, you can provide information alphabetically and give readers indexes and cross-references, but online they can click on an item to see the related information. Consider making reference information available online to make searching faster and more powerful.

When writing reference information, keep in mind that users reading it are typically looking for a single piece of information. For example, they want to know how to use the Gooble function and nothing else. Giving them long narrative text is probably not going to be popular.

Conceptual information

Conceptual information is by far the most difficult information that you'll be asked to write. When you write conceptual information, you provide the "why" behind a feature. Reference information tells you what a function is, and procedural information tells how to use the function. But conceptual information explains under what circumstances feature A is a better choice than feature B. Typically, the introductory chapter in a book is conceptual; it discusses the product and explains what you can do with the product. In some cases, you might also be asked to explain why your product is better than the competition's.

It's difficult to write conceptual information because it requires an understanding beyond what you can learn from looking at a product's interface. For example, based on things you see inside a car, you can't provide the warning that excessive speed is a bad idea—the relationship between speeding and being pulled over by the police is not apparent from looking at a car's dashboard.

Although it's not easy to write, conceptual information is the glue that ties together your documentation. Without it, you'll have nothing but a list of steps. A document that contains nothing but procedures does not provide the reader with significant value—users can probably figure out the steps on their own eventually.

Procedural information

Procedural information consists of steps that tell a user how to perform a task. Because most documentation has a great deal of information about how to use a product, you'll probably spend a lot of time writing procedural information, also called task-oriented information.

The next chapter, "Writing task-oriented information," offers many pointers on how to write effective procedures.

NOTE: Not every product will have task-oriented and reference manuals. Some products often have just a task-oriented user guide that tells the user how to install and use the product. There may be no need for a programmer-level reference.

Dealing with the inevitable schedule changes

Regardless of whether you're writing interface, reference, conceptual, or procedural information, you'll deal with schedule changes on a project sooner or later (most likely sooner). Slipping schedules on a technical documentation project fall into the same category as death and taxes—you can count on them. Now it's a bit of an exaggeration to say that there's *never* been a project during which dates were met, but those projects are in the minority.

Slipping schedules are common because a product's documentation is contingent upon the product itself. A product is going to change as it is developed—particularly a new one—so it's common sense that those changes will affect the documentation. Last-minute fixes, deletions, and additions to a product's functions most likely mean that there will be last-minute changes to the documents as well. A smart doc plan writer knows this and creates a schedule to accommodate changes. A good schedule also contains a "freeze point"—if the documents are to be delivered on time, the product's development must freeze on a particular day. If there are changes beyond the freeze point, the documentation schedule slips accordingly.

Last-minutes changes sometimes mean that there are some trade-offs. For example, to make the deadline for delivery to the printer, you may have to forego a solid edit on a chapter about a function just added to the product. Make these sorts of decisions with the involvement of the client and the documentation team. Even though time (or lack thereof) will play an important part in deciding on any compromises, it's important to keep the book's users in mind. Audience should *always* be a top concern, even when you're in panic mode because of last-minute changes. After all, if late changes to text make a manual less audience-friendly, you undo much of the work you've completed already.

See "The reality of time constraints" on page 135 for more information about making compromises when time is tight.

Experience is the best teacher

Writing a manual for the first time is intimidating. There are many issues you need to remember while writing (audience, task breakdown, and so on), and then there are the slipping schedules, constantly changing information sources, and other issues that need your attention during the documentation process.

If it all sounds like too much to handle, realize that new writers are rarely asked to draw up doc plans, outline manuals, *and* write the documents; senior writers and project managers often handle the planning and project management responsibilities. However, as you become a more experienced writer, your involvement in the management side of projects will most likely increase. Many companies base technical writers' job descriptions on level of experience and the amount of management responsibility the job entails—for example, a level-one writer has a year of experience and has no management responsibilities, whereas a level-three writer has seven or more years of experience and acts as a project lead and as a mentor for level-one writers.

As you work on more and more projects, figuring out how to handle all the elements of a project becomes second nature. Technical writing seminars, a degree from a technical writing program, or this book alone will not make you a good writer. Courses, degree programs, and reference books on writing can help, but working on real-world projects is by far the best way to develop solid technical writing skills.

A vivid imagination can be most useful

Every once in a while, you'll face the Project from Hell (PFH). You'll know it when you see it, but the PFH usually involves getting no information, no software, no access to the developers, and no specs—nothing.

What do you do when you're asked to do the impossible: write about a piece of technology that you haven't seen (and that probably doesn't exist yet)?

One possibility in such desperate circumstances is to extrapolate. Take what little information you do have and make your best guess. If you do guess, make sure that material you write is clearly labeled as a guess—it's particularly important to flag fabricated information before you send your document to technical experts for review.

7 Writing task-oriented information

What's in this chapter

- Elements of a procedure
- Introducing the procedure
- Breaking down a task into steps
- Including the results
- Adding notes, warnings, and cautions
- Using bulleted and numbered lists
- Letting illustrations tell the story
- Organizing information in tables
- Inserting cross-references

Task-oriented writing makes up the bulk of technical documentation—installation manuals, getting started manuals, and user guides, for example. It's also what beginner technical writers usually write first.

When you create an outline for a book, you usually start by putting together a list of tasks that the user will perform. To begin documenting the software, you then try performing those tasks yourself and write down the steps needed to accomplish each task.

However, creating solid task-oriented content requires a lot more than just writing down what the user needs to do. You need to keep your audience in mind at all times, and you also need to consider questions such as the following:

- Does each step represent an action the user takes?
- Are the results of an action clearly explained?
- Would a graphic help explain an action more clearly?

This chapter explains what you need to think about while writing procedures.

NOTE: This chapter is not an exhaustive resource on how to write task-oriented material. Instead, it explains many of the concepts that technical writers must know.

Elements of a procedure

Whether it explains how to use a piece of software or how to bake a cake, task-oriented information generally has the same elements—steps requiring action by the user, information about the results of those actions, and graphics, tables, or notes to clarify what the user does.

Figure 5 shows a sample procedure.

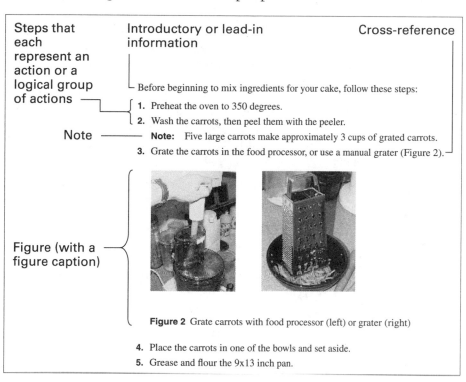

Figure 5: *Sample procedure*

The following sections provide details about how to handle the elements of a procedure.

Introducing the procedure

Before readers dive into a list of steps, they would probably like an idea of what they're about to accomplish. A lead-in sentence or paragraph can do just that.

Often, all that's needed is a simple sentence, like the one shown in Figure 5 on page 93. If you do use sentences for lead-ins, make sure all the lead-ins follow the same grammatical pattern. For example, if one lead-in reads

To print a file, follow these steps:

the next one should read

To delete a file, follow these steps:

and not

You can delete a file by completing the following steps:

If necessary, your lead-in can be a short paragraph, particularly if you need explain any prerequisites for completing the procedure (for example, the user needs to complete another procedure before beginning the one being introduced). A lead-in can also briefly explain how completing a task fits into bigger scheme of using the product.

Breaking down a task into steps

You (or another member of your team) have already written an outline that breaks down the tasks you need to document. When you start writing the documentation, you continue the process of breaking down information—you break down the tasks in the outline into the steps within procedures.

By following a pattern in writing your procedures, you establish a rhythm that makes information easier to retrieve—the readers know what to expect. As Figure 5 on page 93 shows, a step can contain one action or a group of actions. Step 1 in the sample procedure tells the user to just turn on the oven, while Step 2 instructs the user to complete two actions that are closely associated with one another—washing and peeling. In software user manuals, it's common to group selecting a menu and a choice from that menu:

Select the Edit menu, then Copy.

You can also group the actions of typing a command and pressing **Enter**:

Type ftp at the command prompt, then press Enter.

However, don't take grouping to extremes by combining multiple actions, particularly selecting menu choices from two different menus.

Including the results

When writing procedures, it's often helpful for the user if you include the result of each step (particularly in software manuals). For example, if a step tells the user to select a menu choice, the step should include the result of selecting that choice (Figure 6 on page 96).

> **1** Select the **Format** menu, then **Page Layout**, then **Master Page Usage**. The Master Page Usage dialog box is displayed (Figure 21).
>
>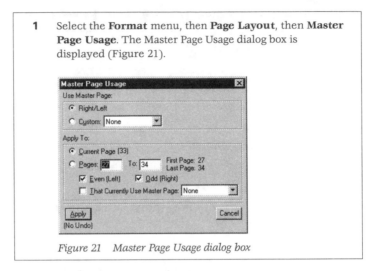
>
> *Figure 21   Master Page Usage dialog box*

Figure 6: *Show the result of an action in a step*

When including the result of an action, don't make it look like a step by placing a number before it. Following the example shown in Figure 6, the result for selecting the menu choices in Step 1 should *not* become the second numbered item in the list of steps:

1. Select the **Format** menu, then **Page Layout,** then **Master Page Usage.**
2. The Master Page Usage dialog box is displayed.

Basically, be sure each numbered list item in a procedure tells the user to perform an action. You could, however, make the result of the action more obvious by putting it on a separate line:

1. Select the **Format** menu, then **Page Layout,** then **Master Page Usage.**
   The Master Page Usage dialog box is displayed.

Adding notes, warnings, and cautions

You can use notes to include information in a procedure that doesn't quite fit into the action/result flow. Use notes for information that needs to be mentioned but that is a bit off-topic. A note contains information that is helpful to the user but that does not involve damage to the product or physical danger. See Figure 5 on page 93 for an example of a note.

If the information concerns possible damage to the product or danger to the user, use caution, warning, and danger paragraphs.

Usually, a caution tells the reader about a potential problem, but it does not involve damage to the product or injury to people.

A warning may explain certain actions that will damage components. In a software manual, a warning may point out that selecting a particular button on the interface will permanently delete a file.

A danger notation indicates the possibility of serious injury or death. Generally, danger notations are found just in hardware documentation, but sometimes software manuals have them, particularly if you're dealing with medical software. For example, for software that calculates medication dosages, a danger notation might point out that entering an incorrect weight for the patient could result in calculating the wrong dosage, which could kill a patient.

Your editor or style guide can give you guidance on when to use and how to format notes, cautions, warnings, and danger notations. Companies often have very specific guidelines about including warnings and dangers to ensure user safety and to reduce product liability.

Be judicious about inserting notes into the text. Overusing notes diminishes their importance and can be an annoyance for readers.

Using bulleted and numbered lists

In general, use a numbered list to denote steps that the user must perform in a particular order, as shown in

Figure 5 on page 93. For lists of items or choices (which do not have a particular order), use a bulleted list, as shown in Figure 7:

Use the following kitchen appliances and hardware while baking your cake and making the icing:

- Oven
- Peeler
- Food processor (or cheese grater)
- Measuring cups and spoons
- Wooden spoons for stirring
- Egg whisk
- Two large mixing bowls
- 9 by 13 pan
- Wire rack
- Oven mitts
- Mixer
- Bowl scrapers

Figure 7: *Bullets don't denote a sequence of steps*

If you have a list of three or more items, consider putting them into a bulleted list instead of listing them in a sentence. For example, if all the items listed in Figure 7 were just part of a sentence, the user would have to wade through the following monster chunk of text:

To make your cake and the icing, use the oven, peeler, food processor (or cheese grater), measuring cups and spoons, wooden spoons, egg whisk, two large mixing bowls, 9 by 13 pan, wire rack, oven mitts, mixer, and bowl scrapers.

The bulleted list communicates the same information in a much clearer way.

Letting illustrations tell the story

Sometimes, no amount of writing can explain something as clearly as a well-chosen screen shot or technical illustration.

In software manuals, it is common practice to show the dialog boxes and windows a user encounters while completing a procedure, as shown in Figure 8:

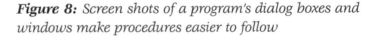

2 From the **Commands** drop-down list, select **New Format.**
The New Format dialog box is displayed (Figure 14).

Figure 14   New Format dialog box

Figure 8: *Screen shots of a program's dialog boxes and windows make procedures easier to follow*

In hardware books, a technical illustration or photograph may depict the action a user is performing in a step, as shown by the photograph with the food processor in Figure 5 on page 93.

If you do insert a graphic, don't assume it is self-explanatory. Include a figure caption to explain its purpose, and if necessary, point out particular items with callouts, which are the lines and bits of text that focus the reader's attention. See the "Hula man" callout in Figure 2 on page 84 and the "Oven" callout in Figure 10 on page 103 for examples.

When deciding whether to include a graphic, consider your audience. For example, if you're telling computer novices about how to open a file, you might want to include a screen shot of the standard Windows Open dialog box. However, for more computer-savvy audiences, such a screen shot would probably not be necessary because the users are accustomed to that standard dialog box. If your product uses a nonstandard Open dialog box, you might want to explain it and show a graphic, though.

If you're working with graphic designers and technical illustrators, be sure to keep them informed about what you need. Also, check the schedule in the doc plan for graphic request cutoff dates. In general, it's a writer's responsibility to take screen shots, but an illustrator can help with drawings and more complex graphics.

See Chapter 8, "A few words about pictures," for more information about creating graphics.

Organizing information in tables

Tables are yet another way to communicate information (particularly information that is repetitive or has a pattern). To ensure that a table's purpose is clear, give it a table caption, and use headers on your columns and rows, if necessary (Figure 9 on page 102).

Table 2 lists the ingredients for the icing.

Table 2 Icing ingredients

Ingredient	Amount
Cream cheese, softened	8 ounces
Butter or margarine, softened	1/2 cup
Confectioners sugar	4 cups
Vanilla extract	1 teaspoon
Chopped pecans (optional)	1 cup

Figure 9: *Tables are useful for information that has a pattern*

Tables can also be helpful in reference material. Many reference manuals have page after page of tables (for example, tables listing commands, their parameters, and their uses).

Inserting cross-references

It's not unusual for multiple figures or tables to appear on a page in a task-oriented document, so it's important that any text that refers to a figure or table be very specific about what's being referenced. The best way to ensure that the user references the correct item is to use a cross-reference.

Placing a cross-reference can be as easy as inserting an identifier for the referenced item in parentheses (Figure 10).

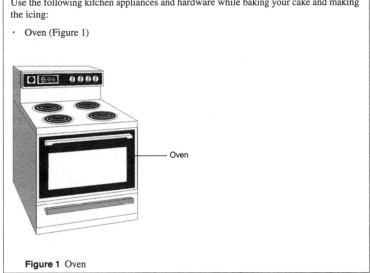

Use the following kitchen appliances and hardware while baking your cake and making the icing:

· Oven (Figure 1)

Figure 1 Oven

Figure 10: *Simple cross-reference*

Figure 5 on page 93 and Figure 9 also show basic uses of cross-references.

Sometimes, it's necessary to include the page number for the referenced item when the item is several pages forward or backward, as shown by the cross-references you just read in the preceding paragraph and by the note in Figure 11:

1. In the second large mixing bowl, beat the eggs with the whisk, then add the oil and the milk.
   Note: Table 1 on page 4 lists the ingredient amounts.

Figure 11: *Cross-reference with page number*

If your text development software has a cross-referencing feature, use it to insert references instead of just typing in the name and page number of the item you're referencing. That way, if the book's pagination's changes (which it undoubtedly will as you add and delete content), the software will automatically update the page numbers in your cross-references. If you manually typed the page numbers for your cross-references, you'd have to verify that every reference's page number was correct before printing the document.

It's not necessary to include the page number if the referenced item is on the same page or a facing page.

NOTE: Sometimes you won't know what pages in a manual will be facing one another until final production, so verifying whether cross-references need page numbers should be part of your production process. See "Page numbers in cross-references" on page 163 for more information.

8 A few words about pictures

What's in this chapter

❖ What sort of graphic should I use?

❖ Understanding graphic file types

❖ Scope of an illustration

❖ Displaying information from your computer screen

❖ Placing graphics in your documents

❖ Hey, I'm a writer, not an artist!

Perhaps the name technical *writer* is a bit misleading. As you create your documents, you'll discover many places where providing information in a graphic will make the document much easier to understand.

Graphics include several different types of illustrations, such as:

- Photographs
- Line art (drawings)

- Screen shots
- Computer-Aided Design (CAD) graphics[1]

The type of information you're trying to convey will determine which graphic element you use.

What sort of graphic should I use?

There's no single correct answer to this question. You generally use screen shots when you want to show a picture of what's being displayed on the computer screen. Photographs are great for showing real-life images or highly detailed information. Line art lets you simplify a piece of machinery and show just the parts that you want the reader to focus on. CAD graphics show exact, detailed views of components.

As always, you need to consider your audience and what they need to know before you decide which type of image to use.

This chapter focuses on the technical details of adding images to your documents, such as the file formats you'll need and how to control the size of your files. For information about designing visual information, consult *The Visual Display of Quantitative Information* by Edward Tufte (ISBN 0961392142).

---

1. Engineers use CAD programs to create precise drawings of items such as circuit boards.

Understanding graphic file types

In the physical world, visual images have lots of different formats—oil paintings, pencil drawings, blueprints, photographs, and so on. When you store images on a computer, however, you only have two types of graphics—vector images and bitmap graphics.

Vector images are images that consist of relatively simple lines and shapes, such as flow charts, diagrams, and line art.

Bitmap images are usually more complex images, such as photographs and screen shots.

Vector images

In mathematics, a *vector* is a line that has a starting point, a length, and a direction (in other words, an arrow). *Vector images* are graphics that are made up of vectors. The vector image file contains a series of mathematical equations that describe the image in the file. A simple image, such as a box, would be defined by instructions something like this: "Start in the upper left corner, with a black pen that's two points wide. Draw a line two inches to the right. From there, draw a line two inches down. From there, draw a line two inches to the left. From there, draw a line two inches up, finishing where you started." That's a vector description, as shown in Figure 12 on page 108.

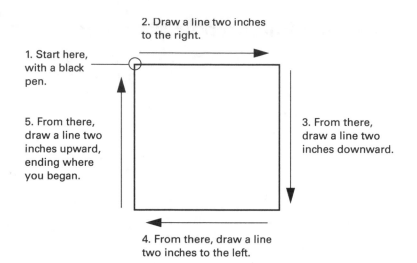

Figure 12: How vector graphic information is stored

The content development tool you're using probably has a drawing tool that will create vector images. Some tools, such as FrameMaker, have feature sets sufficient to create simple flow charts, and Microsoft Word has all that plus a fairly extensive set of useful symbols (such as flow chart boxes). Other tools for this kind of work include MacroMedia FreeHand, Adobe Illustrator, and Microsoft Visio. (Visio is especially good for flow charts.)

Figure 13 shows a typical example of vector art—a Venn diagram, which includes both graphics and text.

If you need to create a diagram or line art, vector images are a good choice. The vector files are quite small because the vector descriptions are compact.

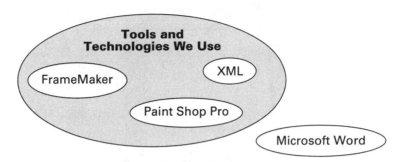

Figure 13: *Venn diagram*

Vector images, however, are a poor choice for photographs and screen shots. In a photograph, you have thousands or millions of different colors, and the image shifts colors unpredictably—a nightmare for a mathematical equation.

Bitmap images

Bitmap graphics handle photographs with ease. A bitmap graphic stores information in a grid. Each square in the grid has a unique address and a color. Unlike the vector image of the box in Figure 12, a bitmap graphic does not have any information about the box; instead, the bitmap stores the image by describing which squares are black (that is, the squares that contain part of the box's outline) and which ones are white (meaning they don't have the box's outline in them). Figure 14 on page 110 shows how bitmap editors describe a picture.

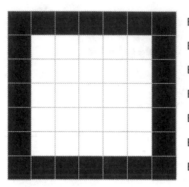

Black, black, black, black, black, black, black.

Black, white, white, white, white, white, black.

Black, white, white, white, white, white, black.

Black, white, white, white, white, white, black.

Black, white, white, white, white, white, black.

Black, white, white, white, white, white, black.

Black, black, black, black, black, black, black.

This is the image, superimposed over the bitmap

This is what the file might look like if it were written in English.

Figure 14: How bitmap images work

This is an inefficient way to draw a box—the vector version is much simpler. But for photographs and screen shots, bitmaps work well because you can't easily describe the image as a series of vectors.

File formats

Graphic files come in many different file formats, such as EPS, BMP, TIFF, PCX, GIF, JPEG, and PNG, just to name a few. Certain file formats are always bitmaps; others are always vector images. Table 2 provides some recommendations for graphic formats.

Table 2: *File formats for graphics*

For this type of graphic	Use this file format
Vector image such as diagram or flow chart	• Encapsulated PostScript (EPS) • Windows Metafile (WMF) • PostScript (PS) • Portable Document Format (PDF) EPS is the standard format for vector images and the most widely accepted.
Photograph	Use Joint Photographic Experts Group (JPEG) format for graphics that need to be as small as possible. JPEG provides good compression for photographic images. Keep in mind, however, that JPEG compression is "lossy"—compressing the file reduces its quality. To preserve every data point, use Tagged Image File Format (TIFF).
Screen shot	• Windows Bitmap (BMP) • Sun Raster Image (RS) • PC Paintbrush (PCX) • Macintosh picture file (PICT) • CompuServe Graphics Interchange Format (GIF) GIF is a good choice because it automatically compresses the graphic. GIF compression is "lossless"—it does not change the quality of the original image.

Scope of an illustration

One important consideration in creating graphics is to give your readers *context*. You want to be sure that readers can identify the main point of the illustration immediately, but you also want them to understand what part of the whole you're highlighting in your graphic.

Consider, for example, Figure 15, which shows the application window for Internet Explorer with the **Help** menu displayed. If you're trying to show readers the location of the **Contents and Index** choice in the **Help** menu, this is probably not the best way to do it.

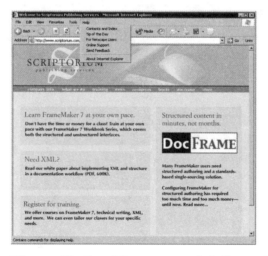

Figure 15: *Providing too much information makes it difficult for readers to locate the item you want them to see*

The image is so large, you had to shrink it to fit on the page, making the lettering in the menu small and hard to read. In Figure 16, the menu choice is easy to see, but the image lacks context; the reader cannot easily tell where in the application the choice is located.

Contents and Index
Tip of the Day
For Netscape Users
Online Support
Send Feedback

About Internet Explorer

Figure 16: *Too little information leaves the reader lost*

Figure 17 shows a better approach. In this image, you show the **Help** menu (including the title at the top) along with just enough of the main window to give the reader an idea of where this menu is displayed in the application.

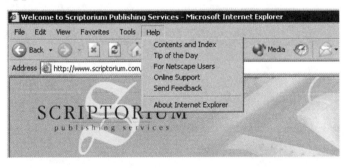

Figure 17: *Now readers can locate the item you want them to see*

For every graphic you include in your document, consider how much context is needed so that the graphic makes sense to the reader.

Displaying information from your computer screen

Most software documentation uses pictures of the interface; these pictures are called screen shots. There are many programs for taking screen shots; for example, Paint Shop Pro and Hijaak Pro are common for Windows machines, and Snapz Pro is a common Macintosh utility. All of these tools let you take a picture of an entire screen, part of the screen, or just an active window. But you don't necessarily need a separate utility to take screen shots. Both Macintosh and Windows systems have built-in screen capture functions. They don't have all the bells and whistles of the commercial versions, but if you just need a couple of screen shots, they're usually good enough.

When colors scheme

When multiple authors are working on a project, all their computers should have the same display scheme to ensure consistency in screen shots. For example, on the Windows operating system, the colors and fonts in dialog boxes vary according to which desktop theme is selected. Double-click the **Display** icon in the Control Panel to see all the options. Authors should also check their screen resolution settings. The resolution affects the size of dialog boxes. Even if a project requires screen shots taken on multiple operating systems, it is a good idea to use the same (or similar) screen resolution to keep the graphics consistent in size.

Ask your editor to add the display and resolution information to the style guide.

Mac screen shots

There are three built-in ways to take screen shots on a Mac. In every case, the file is saved to your hard drive as a PICT file with the name Picture 1, Picture 2, and so on.

- To capture a full-screen image, press **Command + Shift + 3.**

- To capture a rectangular area, press **Command + Shift + 4**, then drag the cursor to define the rectangle. When you release the mouse button, the image is captured and saved.

- To capture a window, press **Command + Shift + 4 + Caps Lock**, then click the window you want to capture.

Windows screen shots

There are two ways to take a screen shot on a Windows machine:

- To capture a full-screen image, press the **Print Screen** key.[1]

- To capture the currently active window, press **Alt + Print Screen.**

When you take a screen shot on a Windows machine, the image is copied to the clipboard. To save the image in its own file, follow these steps:

1   Select the Windows **Start** menu, then **Programs**, then **Accessories,** and then **Paint** to open the Paint utility.[2]

2   Select the **Edit** menu, then **Paste** to paste the image. (You can also press **Control + v.**)

3   Select the **File** menu, then **Save As** to save the file.

---

1. On some systems, you may need to use **Ctrl+Print Screen**, or some other key combination. Consult your computer and operating system documentation.
2. You can also paste the image into other graphics applications.

Placing graphics in your documents

Once you create your illustrations or screen shots, you'll want to put them in your document. Before doing so, consider the following points:

- Whether to link or embed graphics in document files
- Uniformity in the size of graphics

Linking vs. embedding graphics

The procedure for placing graphics varies depending on the application you're using, but most applications give you a choice between *linking* and *embedding* your graphics. When you link a graphic, you store a pointer to the image in the document file. When you embed a graphic, you put a complete copy of the graphic in your document file. Each of these techniques has advantages and disadvantages, which are summarized in Table 3:

Table 3: *Linking vs. embedding graphics*

	Linked graphics	**Embedded graphics**
File size	Because linked graphics contain only a pointer to the original graphic file, the document file remains small.	When you embed the graphic, you insert a copy of the graphic into the document, which increases the file size by at least the size of the graphic.
Portability	To move a document file from one location to another, you must remember to also move all of the linked graphics.	If all graphics are embedded, you can move the document file and be sure that recipients will be able to see all the graphics.
Updating	When you update a graphic, the linked graphics are automatically updated.	When you update the graphic, you must reimport the graphic into the document.

For technical documentation projects, which tend to grow and evolve, it's usually best to use linked graphics. It's worth your time to set up a directory structure and organize your graphics because this lets you prevent "file bloat"—when your files get so big that they are unmanageable. Figure 18 shows an easy way to organize your content and graphic files. In the example, all graphic files are in a subdirectory called "images."

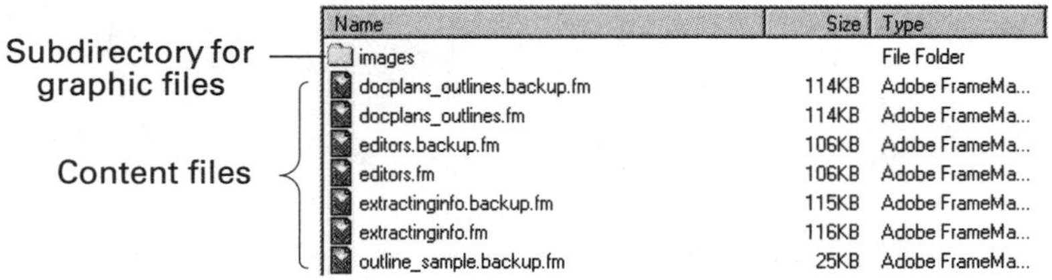

Name	Size	Type
images		File Folder
docplans_outlines.backup.fm	114KB	Adobe FrameMa...
docplans_outlines.fm	114KB	Adobe FrameMa...
editors.backup.fm	106KB	Adobe FrameMa...
editors.fm	106KB	Adobe FrameMa...
extractinginfo.backup.fm	115KB	Adobe FrameMa...
extractinginfo.fm	116KB	Adobe FrameMa...
outline_sample.backup.fm	25KB	Adobe FrameMa...

Subdirectory for graphic files —

Content files —

Figure 18: Directory structure for organizing content and graphic files

Keeping graphics uniform in size

When you place a graphic in a document file, some programs will prompt you to specify the dots per inch (DPI) setting for the graphic. Use one DPI setting consistently—this is especially important if multiple authors are creating content for a book or suite of books. Even if your text processing application doesn't give you the ability to specify a DPI setting for a graphic, you should pick one method for inserting graphics to ensure consistent sizing. Check the user manual or online help for more information.

If you need to modify the size of a graphic after placing it in a file, it's usually best not to adjust the size by selecting the graphic and dragging a corner. Some applications have commands for scaling graphics; scaling preserves the ratio of the graphic's width and height. Figure 19 shows what happens when a graphic's ratios aren't preserved during resizing.

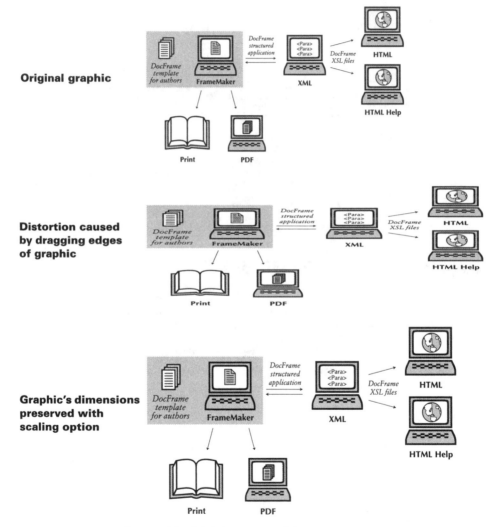

Original graphic

Distortion caused by dragging edges of graphic

Graphic's dimensions preserved with scaling option

Figure 19: *Resizing a graphic*

Hey, I'm a writer, not an artist!

Even if you have no artistic leanings, there is a good chance you will be responsible for creating the graphics in content you are writing. Most technical writers are required to take their own screen shots, and some must draw their own illustrations because no professional illustrators are available to help. It's a good idea to become familiar with the tools for creating simple illustrations. At a minimum, you should be comfortable with the following:

- Taking screen shots
- Creating basic line drawings (such as Figure 13 on page 109)
- Importing illustrations into a document
- Adding callouts to imported illustrations

9 Editors— resistance is futile

What's in this chapter

- ❖ Preventive measures—an editor's early involvement in a project
- ❖ Copy editing vs. technical editing vs. production editing
- ❖ Editing the documentation—what you and the editor can expect
- ❖ Editorial checklists
- ❖ Working with a markup
- ❖ The reality of time constraints

When you think of an editor, you may envision a school-marmish character poised to tear your text to shreds with very sharp red pencils. There are a few editors who have rightfully earned that reputation, but a good technical editor can be a great ally.

Technical editors help you with the organization, tone, and consistency of your document, and they flag spelling and grammatical errors, too. To ensure that manuals are consistent, a technical editor should become involved in a documentation project at its start. For example, by establishing style guidelines early in a project, the editor can save writers lots of work—when all the writers know the rules up front, everyone's text will be much more consistent and will therefore require less rework later on.

Like you, an editor acts as an advocate for the user, ensuring that your material clearly communicates the necessary information at the appropriate audience level. That sort of feedback is invaluable. And, of course, every error an editor catches is one less mistake that the client, reviewers, and users see—and that makes you look better.

NOTE: Some editors may have the title of "technical editor," but all they do is check for spelling and grammar—which is copy editing. True technical editing encompasses checking organization and verifying consistency in tone and presentation. See "Copy editing vs. technical editing vs. production editing" on page 126 for more information about the different types of editing.

This chapter explains the editor's role in the documentation process and what writers and editors can expect from each other on a documentation project.

Preventive measures—an editor's early involvement in a project

The sooner an editor is involved in a documentation project, the better. In particular, there are several tasks an editor should complete in the early phases of a project:

- Reviewing doc plans and manual outlines
- Establishing style guidelines
- Deciding on terminology
- Examining legacy documentation
- Editing early chapters

Reviewing doc plans and manual outlines

Like the technical writers on a project, technical editors should have input on the doc plan, schedules, and manual outlines. It's important to coordinate dates with the editor's (or editorial department's) schedule, particularly because an editor is usually involved in multiple projects, whereas a writer generally has just one.

A schedule should also take an editor's workload into account—it's not reasonable to expect an editor to do a thorough edit of 1,000 pages in less than a week.

NOTE: A decent rule of thumb is that the editor will complete about 10 pages per hour, which means about 400 pages per week. However, that number will fluctuate based on the complexity of the content and the quality of the writing.

Establishing style guidelines

Most documentation departments already have style guidelines in place. These guidelines explain issues such as the following:

- Punctuation
- Capitalization
- Word choice and terminology
- Highlighting
- Acronym use

Many companies develop their own style guides and distribute them to all documentation employees, either in hardcopy or online form. However, some companies follow the guidelines in an established reference such as *The Chicago Manual of Style* and may distribute a smaller, separate document that details any exceptions and additions to the reference's rules. See "Editorial" on page 249 for a list of style references.

Scriptorium Publishing's style guide is available on the web at:

http://www.scriptorium.com/Standards/

If you're doing work for a client, be sure to ask about the client's style guidelines. You and your editor need to ensure that your work matches the client's guidelines.

Alienating your editor in one easy step...

The fastest road to a bad relationship with your editor is trying to impose your personal style preferences on the style guide. Remember, your preferences are just that—preferences—and that's why it's a *style* guide, not a grammar guide. Telling the editor that "only idiots use serial commas" is probably not the best way to build a good working relationship.

The lead editor usually owns the style guide. If you feel strongly that something in the style guide should be changed, try approaching your editor with a suggestion.

Keep in mind, though, that every writer has a set of ingrained style preferences. If the editor changed the style guide to accommodate every new writer, the style guide would change constantly—and frankly, the only thing worse than style guidelines you disagree with is style guidelines that change constantly!

Deciding on terminology

Style guides often contain word usage information that explains the correct way to use particular terms (for example, use *list box* instead of *listbox*) and what terms should be avoided altogether. This usage information, however, may not anticipate terminology specific to the project you're documenting. In that case, writers and editors should work together to draw up a list of terms for the new product. The editor should compile the list of project-specific terms and ensure that all writers have a copy of the list.

Examining legacy documentation

If one of your information sources is legacy documents from the product's previous release, ask the editor to examine those documents for terminology and specific presentation methods. The editor should check with the client about any of the terminology and presentation methods from the previous release that should be preserved.

The editor can also review the legacy documentation to identify any weaknesses in presentation or organization that you should avoid in the new manuals. If you didn't write the earlier version, this is a pain-free way of getting some excellent feedback.

Editing early chapters

It can be very helpful to you for an editor to look at the first chapter or two that you write, particularly if you're a new writer. Early edits can give you an idea of what to avoid as you continue to write, and they can also ensure that all writers on a documentation project are consistent in tone, audience, and presentation.

Copy editing vs. technical editing vs. production editing

Different types of editing include:

- Copy editing—editing text for spelling, adherence to grammar rules and to style guidelines, and general clarity.

- Technical editing—examining the content for organization, presentation, and consistency. Does the text flow logically? Do the text and graphics work together? Are similar types of information presented in the same manner? Also known as developmental editing.

- Production editing—ensuring a book is ready for printing. Are page and line breaks clean? Are headers and footers correct? Are graphics correctly placed?

On documentation projects, a technical editor often does the copy editing and technical editing at the same time. Many companies have production specialists whose primary responsibility is preparing documents for printing. Small companies usually don't have such luxuries, though. See Chapter 11, "Final preparation—production editing," for detailed information about production editing.

Peer reviews: better than no editing at all, but...

Some companies don't hire editors. Instead, writers rely on a peer review system—one writer checks another writer's work. Peer reviews are better than no editing at all, but they cannot provide the same perspective as a technical editor who reviews all the documents for a project.

Because editors generally look at all the books for a product, they gain a deeper view of the entire project that peer reviewers don't. An editor can spot inconsistencies among all the manuals in a documentation suite, but when a peer reviewer looks at one book, the reviewer's comments about consistency are most likely limited to observations about the book reviewed and the book the reviewer is writing. An editor also can evaluate whether all of a product's books create a cohesive unit, but a peer reviewer looking at just one other book probably doesn't have enough information to do this kind of analysis.

It's also difficult for many writers to take an editor's viewpoint, especially considering their primary focus is on writing. As a result, the quality of comments from a peer reviewer are usually not as good as those from an editor (but that's not to say that some writers aren't good editors).

Editing the documentation—what you and the editor can expect

Before an editor actually starts editing your book, you should already have some idea of what will occur during the edit. There are three elements that establish these expectations:

- Doc plan and manual outlines—offer specifics about a document's audience and content. The editor will ensure that the manual's content follows what the doc plan and outline specify. (However, that's not to say that a manual outline is set in stone. Changes to the product in turn cause changes to the outline.)

- Schedule—determines the time an editor has for editing. A short turnaround time may mean that an editor focuses on some issues and disregards others. Decisions of this nature should be made by project managers with input from writers and editors. See "The reality of time constraints" on page 135 for more information.

- Style guide—specifies the guidelines you should follow while writing. If a manual does not show basic adherence to style guidelines, the editor may ask you to correct style issues and resubmit the book for editing, so it's best to follow the rules from the start.

There are also specific things an editor can expect from you and vice versa.

What an editor can expect from you

When it comes time for an editor to edit your manual, you can't give the document to the editor and assume that any problems will be fixed. At most companies, editors don't cut in changes to your work for the following reasons:

- Having an editor mark up a document and return it to you means that you have the final word on changes to your document. If you find a change that you disagree with, you should check with the editor and figure out a compromise. If the editor cuts in changes, you do not have a chance to perform this quality control.

- Seeing the editor's markup and cutting in changes based on the markup will improve your writing skills and adherence to style guidelines. For example, changing every reference to "click" to "select" may be unpleasant, but if the style guide mandates the use of "select," making all of those changes will go a long way toward ensuring you remember to use the correct term in the future.

NOTE: Although writers should incorporate editing changes into their documents as a general rule, editors are usually willing to help out to meet a looming deadline.

The documents you turn over for editing should have the following characteristics:

- Basic adherence to correct spelling, rules of grammar, the style guide, and the template. It's quite understandable and common for a writer not to catch every error or inconsistency, but it is not acceptable for a writer to turn over a document that hasn't been proofed or spell-checked. If a document does not follow the basic rules, many editors will kick it back to you for review, correction, and resubmission to editing. In many companies, the inability (or refusal) to follow style guidelines can lead to reprimands and poor performance appraisals—and possibly the loss of your job.

- Some attention paid to organization. Because a writer gets so close to a document, it's often hard to see how changes in the order of paragraphs or sub-sections would improve the flow of information—the editor's second set of eyes can really help spot areas that could be improved with some reorganization. That being said, however, a writer should not just write material and expect the editor to place it in order.

Basically, paying attention to the style guide, the doc plan, and manual outline—and not expecting an editor to do your scut work—will make an editor quite happy.

"Who does that editor think he is marking up *my* book like that?"

It's best to avoid such a tantrum on technical documentation projects. Remember, just because you're writing a book, it's not really *yours*. It really belongs to the client and ultimately to the reader. If you want to write things according to your own rules, consider taking up creative writing in your spare time.

Having said that, it's fair to ask the editor about any markups you disagree with or don't understand. The editor should be able to tell you the reason for a change, such as, "That doesn't follow the style guide," "That's grammatically incorrect," or "That was unclear to me." If you find that your editor cannot back up a change with a reason, he or she is probably imposing a personal style preference—which good editors avoid.

What you can expect from an editor

You can expect the following from an editor:

- A legible (or semilegible!) markup, preferably in ink instead of pencil. The markup should also use standard diacritical marks (Table 4 on page 132).

Some editors may do things a little differently than others, so after an editor has edited your work for the first time, it's always a good idea for the editor to sit down with you and explain how things are marked up.

- A markup that points out errors and makes commentary but that does not berate you. There shouldn't be color commentary such as "Why did you do that?!?!?!?" or "You fool!"

- A markup that points out positive features of a document. If you've done a good job of presenting information, the editor should note it in the text. Feedback doesn't have to be all negative.

- A list of issues that occur throughout the document (often called a "global list"). Instead of marking recurring errors (such as incorrect capitalization in headings) again and again, many editors will flag an error a few times and then write a comment such as "Global. I won't flag this again in this chapter." The editor keeps a list of these global issues and gives it to you with the markup. Global lists reduce the amount of red ink in a markup, which can improve a markup's legibility. However, an editor should not abuse global lists—a global item such as "book doesn't follow style guidelines; won't mark all deviations" is too broad.

When you receive the global list, it's your responsibility to go through the book's files and make the changes globally. Sometimes, you can use a search-and-replace operation to make global changes, but do so with care. Using search-and-replace with little thought can lead to ugly results, which can take more time to clean up than if you had made the changes individually.

Table 4: *Some standard diacritical marks*

Symbol	Meaning	Example
	Delete	the dialogg box
	Close up	wind ow
	Delete and close up	intra file
/	Change to lowercase	Close the Window.
≡	Change to uppercase	paris
	Change to initial capitals	STOP
STET	Let it stand (that is, don't make the marked change)	log off STET
N	Transpose	dipslay
⊙	Insert period	Insert the card⊙
∧	Insert comma	red, white and blue

Table 4: *Some standard diacritical marks (continued)*

Symbol	Meaning	Example
)	Insert apostrophe	system administrators
the	Insert text	the menu Select **Print**.

Editorial checklists

Some companies give writers editorial checklists to follow before turning over a book to an editor. A sample checklist follows:

Editorial checklist
NOTE: This checklist is a general overview of what editors consider while editing documents. It is not meant as a substitute for the style guide. Please read the style guide and apply its guidelines while you write your document.
Readability and accessibility ___ Steps are in the imperative mood. For example: Select the **Update All** button. ___ Descriptive material is primarily in active voice. (However, don't mangle a sentence to avoid passive voice.) ___ Latin abbreviations (for example, i.e., e.g., and etc.) are not in the text; use the English equivalents instead. ___ Text avoids gender-specific pronouns or constructions such as "he or she." When possible, use plural forms and second person instead.
Capitalization ___ Headings and captions have sentence-style capitalization—not initial caps. ___ Menu names, button names, and the names of other GUI items reflect the capitalization on the interface.

Highlighting
___ Menu names, menu choices, and GUI items are in bold type.
___ First uses of terms are in italic type.
___ Commands are in Courier bold type, and variables in commands are in Courier bold italic type. Command output is in plain Courier type.
Acronym usage
___ On first use in a chapter, an acronym is spelled out and followed by the acronym form in parentheses: local area network (LAN).
Figures and tables
___ Figures and tables have captions.
___ The text contains cross-references to each figure and table.
___ Tables are based on a table format in the template.
___ Tables spanning multiple pages have the table continuation variable in their captions.
___ Figures are inserted according to figure placement rules in the style guide.
Spell-checking
___ Document is spell-checked. Be sure that the Repeated Words, Two in a Row, Straight Quotes, Extra Spaces, Space Before, and Space After options in the Spelling Checker are on.

Working with a markup

When you receive a markup from an editor, spend a bit of time flipping through it. Flag pages that have unclear comments or questions that you need the editor to clarify. As you cut in comments, use a different color pen to check them off so you can keep track of what you've done. You can also write notes to yourself or questions for the editor.

If the editor has marked up something you don't agree with, a reasonable editor will discuss the point with you.

Often, a discussion about a change can lead to a compromise that ultimately serves the product's users better than what either you or the editor wanted. However, don't go crying to an editor about heavy markup focused on style issues—if you had read and followed the style guide while you wrote, that markup wouldn't exist.

The reality of time constraints

Tight schedules sometimes cut into the time available for the editorial process. Although it's always best to anticipate time problems with a schedule that can tolerate some slippage (as mentioned in "Dealing with the inevitable schedule changes" on page 88), you can't always forecast every scheduling problem. If it becomes necessary to make some compromises, be sure to choose ones that have the least impact on the document. Also, if you're working with a documentation team, it's best to get everyone's input about what compromises make the most sense.

The following list describes the features of a good document, starting with the most important. If you need to make sacrifices because of time, start at the bottom of the list.

- Technical accuracy—the book must describe the product correctly. (It's better to have an accurate book that contains spelling errors than an inaccurate document with pristine spelling.)

- Completeness—the book covers all features (or as many as possible).

- Organization—the content flows in a logical manner.

- Index—because the index is often the first thing a user reads in a book, having an index is important, but it's not as important as verifying technical accuracy.

- Grammar and spelling—correct grammar and spelling are essential to clear communication, but if you have to choose between editing a book for grammatical correctness and ensuring it's technically accurate, it's wiser to check for technical accuracy.

- Adherence to style guidelines—application of style guidelines makes text more consistent within a document and across multiple manuals. That being said, however, having all menu names in bold type should be the least of your concerns if time isn't on your side.

- Production editing—ensuring that a document is ready for final production entails checking line breaks, page breaks, running headers, and other items described in Chapter 11, "Final preparation—production editing." While a cleanly formatted document gives an impression of professionalism, nicely formatted text that doesn't reflect the product isn't going to win you kudos from readers.

Just because some of the preceding items are more important than others, don't fall into the trap of thinking the less important items are always disposable. If you find you're consistently making sacrifices, you're not getting enough time or support to create good documentation. You should address the lack of time and resources with your manager. If you're working for a client, keep the client actively involved in making decisions about any compromises, and thoroughly document those discussions—recapping such a meeting via email is a very good idea. (Yes, this is paranoid, but a little paranoia can be a career survival strategy.)

10 Indexing

What's in this chapter

❖ What should I index?

❖ Cross-indexing

❖ Using primary and secondary entries

❖ "See" and "See also" entries

❖ How long should my index be?

❖ Editing your index

❖ Some helpful tips

A complete, user-friendly index is an important component of a professional, finished document. If you think for a moment about how you use technical manuals, it makes sense. You're using a software application that you're familiar with, but you forget how to perform a particular task. To refresh your memory, you turn to the index in the user's guide so you can quickly locate the information you need. But after a few minutes of searching the index with no luck, you end up flipping through the book instead. (Or maybe you give up and call tech support, in which case the book has failed completely.) Instead of looking in the index, locating the page number, and finding the information you need, your quick refresher becomes a time-consuming chore.

The index is one of the most heavily used components of a document. Many readers flip to the index *before* they look at the table of contents. In technical documentation, readers rarely read from cover to cover; instead, they refer to the book only when they are stuck and need help. If readers can't find an index reference to the information they need, they become frustrated.

Professional indexers refer to each entry in the index as an "information access point" (Figure 20).

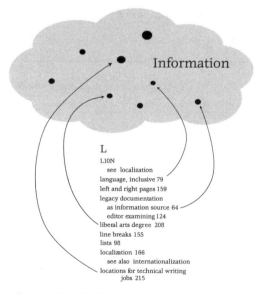

Figure 20: *Information access points*

Thinking of the index in this way can help you identify good index entries. Another way to view an index is to compare it to the table of contents. The table of contents provides readers with a sequential list of topics in the book; it reflects the organizational scheme that the writer of the document chose (which might not be the scheme that the reader would have chosen). However, the index provides a list of topics in alphabetical order, so the readers can locate the information they need using a standard A–Z reference scheme.

Given the critical importance of good indexes to help your readers find information, you may not be surprised that some companies employ professional indexers. However, these companies are in the minority. The task of indexing often falls on the writer, so having basic indexing skills is important. This chapter offers some tips on creating useful indexes.

What should I index?

Let's start with how *not* to create an index. The following sections of a document are usually not indexed:

- The *front matter* of the document, which includes the title page, table of contents, preface, and so on
- The *back matter* of the document, which includes appendices, notes, the glossary, and the index

NOTE: These are general rules. Sometimes, a preface or appendix contains important information that you should index.

Once you begin, don't index every occurrence of a particular word, or you'll end up with index entries that look like this:

chocolate 2, 4, 5, 7, 9, 11, 13, 16, 22, 25, 30

Instead, index only the places where important new information is provided:

chocolate 11, 16, 30

When there are three or more entries, you should break down the entries into subentries so that the reader can scan the list and find the exact topic quickly:

chocolate
>   bittersweet 16
>   milk 11
>   mousse 30

See "Using primary and secondary entries" on page 146 for more information.

Also, don't index the name of every menu choice, dialog box, button, and so on. For example, the following entries are not useful because most readers don't look up topics by their menu location:

Special menu
>   Cross-Reference 15, 18
>   Equations 72–75
>   Page Break 32, 60

Instead, you should generally focus on indexing definitions and tasks:

C

cross-reference, inserting

E

equation, inserting

I

    inserting
      cross-reference
      equation
      page break

P

    page break, inserting

Indexing definitions

Throughout technical documentation, you'll see information like this:

Extensible Markup Language (XML) defines a standard for storing structured content in text files.

and then later in the document:

XML shows much promise.

The first sentence is useful. It provides information about what XML is, so a reader could look up that page and get the definition. You should provide an index entry for that sentence. But the second example is much less useful. It doesn't contain any concrete information, so it's probably not worth indexing. (And maybe the editor should edit it out, but that's another issue.)

If the defined term is an acronym, as in the XML example, you should also include an index entry for the acronym's spelled-out meaning:

Extensible Markup Language
    see XML

The "see" reference helps readers unfamiliar with the acronym find the information they need, and you're saved the extra work of indexing the information under both the acronym and its meaning (refer to ""See" and "See also" entries" on page 147 for more information).

If the document's definitions are in a glossary, you don't need to index them. Because the glossary is in alphabetical order, the reader can easily find the terms there without referring to an index.

Indexing tasks

Readers use procedural information to figure out how to accomplish a task. Therefore, you should provide them with a way to access that information from the index. If you have a section in your document about printing, provide an index entry for "printing." If you have a section about replacing toner cartridges, try "replacing toner," "toner, replacing," and perhaps "ink, replacing."

However, before you index a task, consider your audience. For example, a section of your document discusses querying a database. If the book's audience consists of database experts, the entries "querying database" and "database, querying" should suffice. But if your audience contains users generally unfamiliar with databases, consider adding "accessing database."

Analyze who your readers are and try to determine how they might look up terms (see "Audience, audience, audience" on page 76 for information about audience analysis). For some procedures, you may need to include several synonyms for the same operation ("closing application," "exiting application," and "quitting application").

When indexing a task, be sure to index not only the action ("opening file," "closing file," and "printing file") but also the item involved in the action:

file

opening 10, 34
closing 15
printing 55

Dual entries such as these reflect a good feature of indexing—cross-indexing.

Cross-indexing

Not everyone thinks exactly alike, and your index should anticipate this. By including variations of an entry, you can make sure that different readers find the information they need. (Professional indexers refer to indexes with cross-indexed entries as *permuted indexes*.) For example, suppose there is a paragraph about displaying the text that creates a document's running footer. To cover all your bases, your index could contain the following entries for that one paragraph:

displaying running footer text
footer, displaying text
running footer, displaying text
text, displaying running footer

You will need to include these variations every time you index the topic in the document. It's easy to forget all the variations you're using, so you should write them down for easy reference.

Using primary and secondary entries

A good way to organize the content of your index is to use primary and secondary entries (Figure 21).

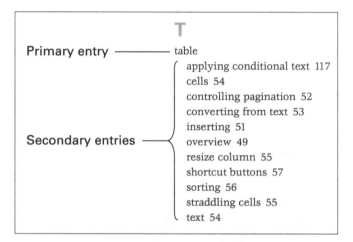

Figure 21: *Primary and secondary index entries*

If the entries about tables weren't combined under the primary "table" entry, the reader would have to contend with something like this:

    table, applying conditional text
    table, cells
    table, controlling pagination
    table, converting from text
    ...

"Collapsing" all the table entries into secondary entries makes finding them much easier.

You can also create third-level entries, but it's probably best to avoid going further than that unless your book is extraordinarily long (more than 500 pages or so).

"See" and "See also" entries

In addition to cross-indexing, you can use "see" entries to be sure your index is useful to as many as people as possible. For example, some readers may look up "changing," but others may look up "modifying." Instead of creating full entries for both terms, you could create a "see" entry that points the reader from "modifying" to "changing" (Figure 22). That way, your index works for people who look for either term, and it isn't padded with duplicate entries.

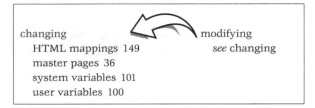

changing
    HTML mappings 149
    master pages 36
    system variables 101
    user variables 100

modifying
    see changing

Figure 22: *"See" entry refers readers to the information they need without duplicating index entries*

A "see also" entry is similar to a "see" entry, but you use it when you have related entries in two locations (Figure 23).

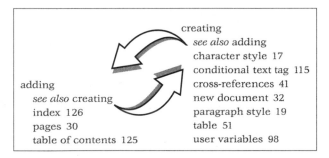

adding
    see also creating
    index 126
    pages 30
    table of contents 125

creating
    see also adding
    character style 17
    conditional text tag 115
    cross-references 41
    new document 32
    paragraph style 19
    table 51
    user variables 98

Figure 23: *"See also" entries refer readers to related entries*

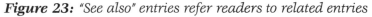

How long should my index be?

The answer, as always, is "it depends." If the material is very dense and complex, the index might be longer than usual. If the information is very "fluffy," the index might be shorter. In technical documentation, however, the rule of thumb is that the index should contain approximately one page of two-column index entries for every 20 pages of documentation. Thus, a 100-page book would have about a 5-page index with two columns of entries (see the index of this book for an example of a two-column format).

Indexing with your text processing program

To create index entries in some text processing tools (including FrameMaker and Microsoft Word), you insert bits of code, or markers, into the document files. When you generate the index file, the software scans the document files for the index markers. The index file that is generated includes entries based on what you typed in the markers and automatically inserts page numbers based on the markers' locations—you don't have to worry about typing the page numbers for entries, and the software alphabetizes the entries for you.

Some applications require that you use special symbols or code to create index markers; your index will be a mess if you don't understand this coding. Also, there may be a third-party software add-on for your text processing software that makes indexing even easier (for example, IXgen for FrameMaker, available at www.fsatools.com).

Editing your index

Once you've tagged your index entries, you're only half done. Most indexes also require quite a bit of revision. To make the process easier, you may want to index a chapter or two of the document, generate your index, and then review what you have done so far.

Here are some problems to look for:

- Cross-indexing inconsistencies:

  displaying running footer text 45, 62
  footer, displaying text 45, 62
  running footer, displaying text 45
  text, displaying running footer 45, 62

  In this example, the second page reference is missing from "running footer, displaying text."

  You should also make sure that entries are cross-posted under both the action and the item receiving the action:

  file
     closing 18
     opening 14
     printing 60

  Confirm that you also have the entries "closing file," "opening file," and "printing file."

- Page-range errors:

  querying databases 32–??

  In this case, the page range start and end tags don't match.

NOTE: This example is taken from FrameMaker, where range errors are shown by double question marks. If you're indexing in a different application, the problem may look a little different.

- Not enough detail in a broad page range:

  printing file 62–80

A page range this long is useless. Go back to those pages to see where you can break down the topic into secondary entries:

printing file
    crop marks, including 78
    ...
    print preview 62
    setup 64

Be sure to include your new subentries as their own main entries as well.

- Too much detail in subentries:

variables
    creating 15
    deleting 16
    editing 15
    inserting in code 15

In this case, you should probably collapse the entry into a range:

variables 15–16

As usual, confirm that entries exist for "creating variables," "deleting variables," and so on.

- Missing index entries referenced by other entries. For every "see" and "see also" reference, be sure that the entry you're pointing to exists.

- Parallelism:

file
    closing 32
    open 23
    saving 30

In this example, "open" should be changed to "opening" to ensure consistency with the other subentries.

Similarly, you should decide whether to refer to items as singular or plural ("file, opening" vs. "files, opening") and be consistent among entries. Also look for consistency in the format and placement of "see" and "see also" entries.

You should review and revise your index several times before submitting it to the technical editor or peer reviewer (and don't be surprised if the editor flags a bunch of errors and inconsistencies you missed).

Some helpful tips

You may feel a little overwhelmed the first few times you index your documents. Here are a few tips to make the process easier:

- Before you start indexing, be sure you understand how to insert markers correctly. Reading over the indexing section of your text processing or desktop publishing program's manual will probably save you headaches later.

- Don't put off editing your index until the end. Frequently regenerating your index and making the necessary revisions is the best way to make indexing more manageable.

- Place your index marker as close to the referenced topic as possible, usually at the beginning of the heading or sentence where the reference occurs. If the marker is too far away from its intended reference, the page numbers may be incorrect in the index. You also want to ensure that if the topic is moved or deleted, the corresponding index marker is automatically moved or deleted as well.

- Be careful indexing symbols (such as \ and ;). Some symbols have special meanings in a program's indexing utility. Consult the program's manual for information on indexing symbols correctly.

- If you make changes to an index, be sure you change the markers themselves, not the entry in the generated index. Changes made directly to the index are usually lost when you regenerate the index.

11 Final preparation— production editing

What's in this chapter

- ❖ What the production editor expects from you
- ❖ What to expect from a production editor
- ❖ The production edit
- ❖ Production checklists
- ❖ Preparing final output

Production editing, or preparing a book to be sent to the printer, is an important responsibility. Writers and editors are concerned with the content of the book— completeness, organization, grammar, punctuation, and the like. The production editor focuses on formatting issues and the appearance of the book.

NOTE: This chapter focuses on printed documents and books converted to PDF format[1]—not online help or HTML output.

By the time a document reaches the production cycle, it's been written, edited, and proofed. As a result, you can expect that the content is clean. Now, it's time for someone to focus on how that content is presented.

Just as the technical editor ensures that the book follows the style guidelines, the production editor ensures that the book follows the formatting guidelines. For most applications, this means using the specified template and applying the correct styles to various elements in the book. The production editor also looks at page elements, such as headers, footers, and page numbers, to ensure that they are placed consistently and contain the correct information throughout the book.

Like copy editing, production editing is an invisible art. When a book is free of typographical errors, readers don't notice—you'll rarely hear someone say, "Wow, that book was great. It didn't have any typos in it." But readers notice typos, which make a book look unprofessional and amateurish. Similarly, production errors are noticed only when they occur, not when they are absent. Typical production problems that an editor looks for include the following:

- Inconsistent formatting of elements
- Strange page and line breaks
- Incorrect headers and footers

---

1. See page 53 for an explanation of PDF files.

- Inconsistent graphic sizing
- Typesetting mistakes, such as double hyphens instead of em dashes
- Pagination problems

Some documentation teams don't have a dedicated production editor, so writers do their own production editing. Again, there are similarities to editing your own work. It's very difficult to step back from a book you've just written and look at the formatting dispassionately. Handing off your book to a production editor saves you time and generally means that your final product will be much improved over what you could do on your own.

What does it take to be a production editor?

In many companies, the production editor is on the same level as an editorial assistant—the position is considered an entry-level training position. This is unfortunate, because the skills required for a successful production editor are quite extensive. To provide a useful production edit, you need extensive tools knowledge, and you need to know how to "cheat" to make the software do what you need it to. Once an entry-level person learns how to do a really solid production edit, that person is often promoted into a writing or editing position, so the training process starts over with another newbie.

What the production editor expects from you

Your relationship with your production editor will be greatly improved if you follow the submission guidelines that your production editor provides. Generally, these include the following:

- Documents should follow the approved template with few or no formatting overrides.

- If you created custom styles, document where and why they were created and submit this information to the production editor when you turn over the book.

- The document should be complete. A production editor cannot work with books that are submitted a chapter at a time.

- Graphics should be set up using the guidelines established by your company.

What to expect from a production editor

Production editors should focus on the formatting of your document, not on the content. The production editor generally has leeway to correct hyphenation and pagination problems but should not change the text. A good production editor who spots a possible problem in your content will consult with you before making any content changes.

The production edit

Where does production editing begin? The first step is usually to open all the files in the book and apply the styles in the company template, thus removing all formatting overrides in the files.

The next step is to check the formatting catalogs for additions. If you're working with a production editor, expect to have any new styles challenged at this point. In a large company with a stable template, it's unlikely that you'll be allowed to add styles. In a smaller company where the template is still under development, you might be allowed to keep your custom styles, and the production editor may ask you for more information about the styles to figure out whether they should be made available to all writers as part of the standard template.

NOTE: Some production editors greatly enjoy blowing away renegade styles; others are irritated by them. If you do have a production editor and want to maintain a good working relationship, don't create your own styles. Adhere to the template just as you adhere to the style guide.

In addition to checking compliance with the template, the production editor checks each page of the document for several items. The next few sections discuss some of the issues.

Hyphenation and bad line breaks

The production editor looks for hyphenation that might confuse the reader. These include some classics such as:

mole-ster

anal-ysis

When hyphenation like this occurs, the production editor will adjust the hyphenation points in a word to correct the problem. The result would be:

moles-ter

analy-sis

In addition to adjusting hyphenation to prevent misunderstanding, the production editor also looks at the end of each paragraph. Leaving a word fragment on the last line of a paragraph is disconcerting to the reader.

In Figure 24, the problem was corrected by forcing the word "the" in the second line to the third line. The result is a much more balanced paragraph.

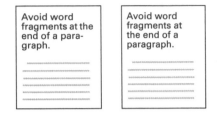

Figure 24: *Avoid word fragments on their own line*

Many text preparation and word processing programs have automatic hyphenation that works really well—most of the time. In some cases, words will hyphenate across pages, or just a few syllables of a word will dangle at the end of a paragraph. Eliminate hyphenation of this nature.

Also, if there is a paragraph with a lot of hyphenation at the ends of lines, eliminating the hyphenation in one line may eliminate hyphenation of the next line. In general, it's best to avoid having two concurrent lines of text both ending with a hyphen. Some publishing packages let you prevent this by setting a preference or an attribute for the paragraph styles.

Page breaks and copyfitting

You may not have ever thought about this, but when you encounter a page in a book that's only half-filled with text, you generally assume that you're at the end of a chapter or section. If you turn the page and discover that the chapter continues (perhaps with a big figure that didn't fit in the space on the previous page), this causes some confusion. Checking the copyfitting (how much text is on each page) can help you avoid these problems. As shown in Figure 25, you can move text from the right page over to the bottom of the left page to get rid of the empty space at the bottom of the left page.

Figure 25: *Large illustrations often cause copyfitting problems*

You also want to avoid having a heading at the bottom of the page and the introductory paragraph at the top of the next page. In most applications, you can eliminate this problem by requiring that a heading style always stay on the same page as the following paragraph.

NOTE: Do not press the **Enter** or **Return** key to create page breaks. Inserting blank lines in this manner is unnecessary in most publishing applications. Check the user guide or online help for your program to determine the correct way to force page breaks.

Other bad page breaks include a figure or table being on one page while its caption is on the next. This looks silly, but worse, it can also cause navigation problems. Because many documents' cross-references point to specific figure and table numbers, the caption needs to be on the same page as the referenced item.

> ## Writing + fixing page breaks = waste of time
>
> While writing content, don't worry about how pages are breaking. If you adjust pagination as you write, your efforts will be in vain—product changes and edits will mean additions and deletions to text, which in turn affect page breaks. Wait until you are done writing and cutting in changes (in other words, wait for the production edit) to adjust pagination.
>
> In some writing groups, setting page breaks counts as a template override, which will get you in trouble with your production editor. Keep that in mind when you're deciding whether to "tweak" your pagination.

Widows and orphans

In addition to paying attention to how paragraphs break across pages, you also need to watch how lines of text break across pages. Sometimes, the last few words of a sentence end up at the top of the page. Such a fragment is called a *widow* (Figure 26).

Figure 26: *Widows are a few words at the top of a page*

The opposite problem—the first line of a paragraph alone at the bottom of a page—is called an *orphan.*

Most word processors and desktop publishing packages let you set up your paragraph styles to prevent widows and orphans. Nonetheless, you should check your pages to ensure that none sneaked into the book.

Right/left pagination

In a printed book, a chapter title page usually falls on a right page. That means that odd-numbered pages need to be on the right, and even-numbered pages need to be on the left. If a chapter's content ends on an odd page, add a blank even page (Figure 27). Some applications automatically add blank pages to make a chapter end on an even page.

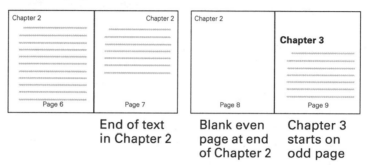

End of text in Chapter 2   Blank even page at end of Chapter 2   Chapter 3 starts on odd page

Figure 27: *Use a blank page to make a chapter end on an even page*

Some companies require that blank pages be marked with "This page intentionally left blank." Check with your editor and template designer to find out whether blank pages need special treatment in your manual.

NOTE: In a PDF file, you don't need to worry about odd/even pagination because the concept of right and left pages doesn't apply. Therefore, blank pages generally aren't necessary in PDF output.

Running headers, running footers, and pagination

Running headers and *running footers* are the bits of text at the tops and bottoms of pages that contain information such as manual title, chapter name, subsection name, and page number—just look at the top and bottom of this page for an example. Before the advent of word processing software, checking footers and headers was crucial because they were inserted manually by the typesetter, who might put the wrong chapter title in a running header accidentally. Many software packages today handle this chore automatically, particularly if you have a good template, but that doesn't mean you should blindly trust the software (mostly because people using software do make mistakes). While performing a production edit, make sure that the headers and footers reflect the correct information. For instance, a common mistake is to use the chapter's running header in an appendix, which results in something like "Chapter A" instead of "Appendix A."

Just as you shouldn't completely rely on your software to generate headers and footers correctly every single time, you shouldn't assume that the software will always correctly number your pages. In some text processing programs, you can reset pagination at the start of a chapter, and if that happens accidentally, you'll end up with a second Page 1 in the middle of your book, and you probably don't want that.

So, although software really can help you focus on creating content by handling many cumbersome formatting tasks, you should always check during a production edit that the software did what you thought it would do.

Consistency in presentation of tables and figures

Even though much of a production edit focuses on text issues, you should also look at figures and tables to ensure they are consistently presented.

Figures should have the same amount of white space around them, and borders around them should be consistent. Tables should have similar shading and ruling. You may have two or three standard formats for your tables, and they should be used consistently. Because you can define table types in some software, it's easier today to ensure that tables look the same—just ensure that tables use the same table format.

When reviewing tables and figures, also ensure that they have numbers and captions (if your company's style mandates their usage). Also verify that cross-references to figures and tables are accurate—users get very frustrated when a step refers to the wrong figure.

Page numbers in cross-references

Once pagination is stable, check to see if cross-references to tables, figures, and the like need page numbers. Generally, if a cross-reference is on the same or facing page as the referenced item, including the page number in the reference is unnecessary (Figure 28).

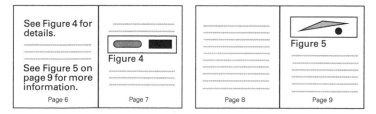

Figure 28: *Cross-references to items on the same or facing page should not include page numbers*

Production checklists

Regardless of what software your company uses to develop content, there are several general things you need to look for during a production edit.

To help writers and production editors keep track of what they should check during a production edit, many companies create production checklists similar to the following:

Production checklist
NOTE: This checklist gives you a general idea of what production editors consider while editing documents.
Page-level editing   ___ Are the correct styles (paragraph, character, table) used?   ___ Are headers and footers correct?   ___ Are heading styles sequential (heading 1 is followed by heading 2, not heading 3)?   ___ Are registration and copyright symbols superscripted?   ___ Are quotation marks curly instead of straight (except in computer code)?   ___ Have graphics been scaled and placed uniformly, and do they have captions?   ___ Do table titles and headings continue when tables breaks across pages?   ___ Did you prevent hyphenation of words across two pages?   ___ Is computer code in Courier font?   ___ Are at least two bullets or steps together before a page break? (Avoid having one bullet or step on a line by itself before a page break.)   ___ Did you prevent a page or line break between words containing hyphens, forward or backslashes, and em or en dashes?   ___ Did you prevent a short word (four or fewer letters) from being on a line by itself?

Page-level editing (continued)

___ Do cross-references to items not on the same or facing page include the page number?

___ Did you prevent hyphenation that would create 2-letter prefixes or suffixes? (Some software programs contain prefix and suffix hyphenation settings.)

___ In the table of contents and index, are TOC and index entry numbers on the same line as the entry? (Prevent the number from being on a line by itself when possible.)

Book-level editing

___ Do chapters start on the right-hand page (if a double-sided book)?

___ Has the book cover information been updated (title, author, version, and logo)?

___ Has the front matter been updated (copyright date and text, ISBN, credits, and part number)?

___ Are autonumbered styles numbering correctly (throughout the book or restarting in each chapter)?

___ Are index entry ranges missing (the start or end range is denoted by question marks)?

___ Do the table of contents and index have sufficient entry levels (up to three)?

Preparing final output

After you complete a production edit, it's time to get the files ready for the printer. Usually, this means creating a PostScript or PDF file. Be sure to talk to your printer about the specs for output files. Refer to the documentation that came with your content development software for information about creating files for printing.

12 Avoiding international irritation

What's in this chapter

- ❖ Some basic definitions
- ❖ Did we mention audience?
- ❖ The myth about images
- ❖ Designs that won't hurt you
- ❖ Think globally, act locally

As a technical writer, there's a good chance you'll write documentation that will be translated for other markets. The basic practices for good technical communication still apply when you write for an international audience. However, some practices that seem benign to source developers ("source" meaning the original language in which a document is written) can cause many problems with the redesign that occurs as part of the translation process. In addition, some writing techniques that may work well for informal writing or for nontechnical writing are wholly unsuitable when you write for an international audience.

This chapter provides some caveats about writing and designing documentation for international audiences. Although these guidelines are not exhaustive and cannot eliminate all redesign issues during translation, they can help minimize rewriting and redesign.

Some basic definitions

The term *translation* doesn't adequately address all of the work that occurs when a product is redesigned for another market. The process involves far more than just rewriting the content in a different language. To encompass a broader set of requirements, the term *localization* was coined. Localization is the process of redesigning a product for a specific international market or *locale*. It incorporates translation, file preparation, reformatting, visual proofing, and additional tasks that ensure that the proper characters appear in the text, that the page layout works for all languages and locales, and that specific cultural references or designs are not included in the localized product.

Localization is often abbreviated L10N (which means, literally, L, then ten letters, then N). Ideally, localization involves what happens after a product is completed in its source language to make the product ready for other markets. More and more, this sequence is becoming skewed, where the localization process begins before the final version of the source product has been completed. Such scenarios often result in higher localization costs, lower product quality for localized versions, and a whole lot of frustration for everyone involved. Many localization vendors spend a significant amount of time educating clients about the localization process to avoid these problems.

One way to avoid problems during localization is to *internationalize* the product. *Internationalization* (or I18N—care to guess why?) is the practice of designing a product to be as culturally neutral as possible, accounting for issues such as language, design conventions, and tool limitations. By internationalizing your documentation, you can lessen the frustration that you may otherwise encounter after you finish a source product and send it through the localization process.

Did we mention audience?

As mentioned earlier in this book, technical writers must have a clear vision of the audience for whom a document is intended (see "Audience, audience, audience" on page 76). When you write for an international audience, this advice is even more crucial. You need to remember that not all countries have the same level of technical sophistication, employ the same methods for education, or have the same expectations of technical documentation as the primary audience you're writing for.

In addition to differences in exposure to technology, people from different locales have different social mores and a different body of cultural knowledge. What might make sense to someone raised in the U.S. may have little relevance to someone from another locale. In the same respect, what passes for humor in one culture may be unfunny or even offensive in another culture.

So technical writers need to be aware of the cultural context in which their work may be viewed and write appropriately for the situation.

Also, many of the common phrases we use to describe different geographic regions reflect our bias. We refer to the southwestern part of Asia as "the Middle East" and the southeastern part as "the Far East." (Ironically, from the United States, you usually fly west to get to the Far East.) Geography is relative to location, so every locale has its own "Middle East." Using more precise geographic descriptions can help you avoid the relativistic blinders that lead to such references. You also need to be aware of ethnocentric or U.S.-centric comments that sometimes creep into marketing statements or product descriptions. Although a technology may help make a U.S. business "more competitive against the Japanese," it can also work in the reverse. In a global market, writers need to adopt a global perspective.

Language is an eight-letter word

Okay, so maybe an eight-letter word doesn't sound all that offensive. However, language creates a dilemma for content developers. Language, according to George Lakoff and Mark Johnson in *Metaphors We Live By* (ISBN 0226468011), is inherently metaphorical. As such, it's often most descriptive when it is used figuratively. And the authors note just how often our everyday speech relies on figurative speech for comprehension.

The problem is that different cultures have different conventions, mores, and expectations about what is acceptable and what is not. For example, speakers of

U.S. English would have no difficulty understanding the allusion in the heading of this section. An eight-letter word somehow matches a four-letter word in its offensiveness. However, to someone who speaks another language, the phrase "four-letter word" may have no significance.

In the U.S., we use an idiom to describe an event that occurs infrequently: *once in a blue moon*. Although most U.S. English speakers understand the idiom, very few actually know what a blue moon is. Nonetheless, idioms still manage to convey meaning within a locale. Outside a locale, the idiom often doesn't work. For example, in Italy (which has a large population of Catholics) the same meaning is conveyed by the phrase, *every time the Pope dies*. Compare these to the equivalent in Ecuadoran Spanish, *every time there is an eclipse*. Each phrase is comprehensible if in context, but the immediacy of the meaning is not the same between locales. Imagine how the following U.S. English idioms might sound to non-U.S. ears:

- I got it from the horse's mouth.
- He kicked the bucket.
- It's like finding a needle in a haystack.
- That fellow is two bricks short of a load (or "dumber than a sack of hammers," "a taco short of a combination plate," and so on).

Idioms are based in regional and national dialects, and as such, they create problems for translators and for audiences.

More on mores

Beyond the metaphoric or figurative aspects of language, different cultures *expect* varying degrees of formality. In the U.S., audiences are accustomed to informal social interactions, using first names with people we barely know (if we know them at all), and interjecting humor into subjects that are normally dry and uninspiring. Other cultures are far more formal. Using first names in the workplace may be considered acceptable to U.S. employees, but such behavior in many European and Asian countries would be overly familiar and disrespectful. In many locales, coworkers are expected to refer to people by surnames and formal titles.

Hand-in-hand with the level of formality is the appropriateness of humor. Audiences in many non-U.S. locales expect technical subjects to be, well, technical and consider a serious tone to be a display of respect on the writer's part. Humor in such publications might strike these audiences as condescending or patronizing. While U.S. audiences have no angst about purchasing a book with *for Dummies* or *Complete Idiot* in the title, audiences elsewhere refuse to buy a book that suggests they are less than competent.

Some common-sense rules for international writing

Being aware of the potential for problems is only half the battle, so here are some methods you can use to avoid problems. Some of these ideas have already been covered in this book.

- Write using clear, unambiguous, and consistent terminology.

Use one term for a concept and focus on clarity and simplicity. As long as the meaning of a sentence is clear, a translator can work with it. Using a term as more than one part of speech (such as "press Enter" and "Enter your password") is likely to create confusion, so use a particular term for one meaning and as one part of speech (that is, always a noun or always a verb). Also, use common, basic terms that get your point across. Using an extensive vocabulary and academic-sounding prose is not likely to help you communicate more effectively, and the more subtle you try to get with language, the more likely those subtleties will be lost on your audience.

Really not good: If the printing device fails to initiate the startup sequence within an ordinate amount of time, verify whether it is necessary to reestablish any power or bidirectional connectivity.

Better: If the printer does not start immediately, confirm that it is plugged in and connected to the computer.

- Write using simple sentence constructions.

In creative writing, you write for artistic effect, and sometimes that means complexity and ambiguity encroach on your writing style. The variance between stark imagery and ambiguity is part of the effect a creative writer is trying to evoke, so using complex sentence structures and inverted syntax aids in that endeavor. In technical communication, such devices get in the way of clear communication. Simple sentence structures provide fewer chances for translators to misunderstand your point. Also, make sure you include all optional parts of speech, including relative pronouns. These syntactic clues give translators more on which to base their decisions.

Not so good: If, when the application finishes saving the file, you want to close the application, you can do so by selecting **Exit** from the **File** menu.

Better: To close the application after it finishes saving the file, select the **File** menu, then **Exit.**

- Don't use culture-specific metaphors, idioms, or allusions.

  As noted before, different cultures share different worldly experiences. Metaphors, idioms, and allusions are all part of that experience, so these linguistic devices often do not translate well into other languages or local conventions. Focus as much as possible on the literal meaning of your words to make your point. For example, write "discover the cause of a problem" instead of "ferret out the problem."

- Don't refer to celebrities, sports, or politics.

  Our celebrities aren't necessarily well known elsewhere, and references to sports are lost in cultures where the athletic activities differ. For example, to a U.K. resident, American baseball looks a lot like a game called rounders. And the term "football" is used for at least three very different games internationally.

  What people watch at the movie theater also depends on who distributes films in their region. Depending on cultural constraints imposed by the government, local tastes, or other factors, what appears on film or television and how nonnative

films or programs may be promoted can vary considerably. (For example, in Beijing, the film *Nixon* was screened as *Nixon: The Big Liar.*)

Politics is simply a good topic to avoid because your audience will likely include people of widely divergent political beliefs.

- Don't interject humor.

  Like metaphors, idioms, and allusions, humor is also dependent on culture, and what people in one culture find funny may simply be ludicrous or downright offensive in another culture.

- Don't use first names or gender references in examples.

  In many cultures, people use only titles or surnames in the workplace. In countries where surnames can reveal caste or religious affiliation, some combinations of names could offend readers. Instead, use titles or roles in examples. Using plural terms instead of singular can also help eliminate gender references.

  Not good
  Kathy, Bob, and Sharmilla
  Still risky
  Ms. Jones, Mr. Renshaw, and Dr. Srinivas
  Better
  The system administrator, the database administrator, and the senior network engineer

The myth about images

"A picture is worth a thousand words" may hold true in some situations, but words are far easier to translate. Many writers are unaware of the part culture plays in developing visual literacy. How audiences interpret graphic images depends on their native environments.

Take, for example, the types of icons that have appeared on email applications (Figure 29).

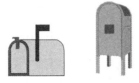

Figure 29: *U.S.-centric postal images*

Outside the U.S., these icons are not so easy to identify. Most Europeans have never encountered mailboxes like these. For some audiences, the closest match they can find is a railroad tunnel.

Much ado about taboo

Another important fact to consider is that some cultures have taboos about how people are represented in images, especially when those images represent people of different genders. Some countries have guidelines about appropriate dress and behavior for people of different genders. These restrictions can include the following:

- Western-style dress for women

- Length of facial hair for men

- Physical position of women and men when interacting

- Position of hands or feet in relation to the perspective of the audience

For example, placing the soles of feet or the palms of hands toward the audience is construed in some cultures as an insult. Because the left hand is considered unclean in many countries, displaying it in images can be inappropriate, depending on the context. If the audience may be offended by the representation, you need to respect that cultural perspective—even if you consider the conventions to be based on sexist or biased religious perspectives. Otherwise, you risk alienating that audience toward the product you are documenting.

Keep in mind that color can also communicate messages, some of which may be inadvertently interjected into your documentation if you are not aware of them. Whereas a hand with the palm facing toward the viewer can be offensive in some cultures, a red hand (often used to indicate critical warnings or cautions) can signify death. Some colors that Westerners would consider innocuous may hold symbolic or even political relevance in some cultures. (For example, in Japan, the color white signifies mourning.) Although it's not desirable to eliminate all color from documentation, you must be aware of the potential for communicating inappropriate messages.

Screening your graphics

You may never know what in an image could offend or alienate an audience from another culture. To make the most of your internationalization efforts, work with your localization provider to ensure that the content you deliver (graphic and otherwise) is appropriate for all audiences. Translators are typically natives of the locales they translate for, and they take pains to stay aware of the mores for their respective cultures. Requesting an analysis of the graphic content (or all content) from your localization provider before you begin writing can help ensure that the documentation you deliver meets the needs of all audiences without offending any of them.

Designs that won't hurt you

The biggest impact on an localization process can occur because of design decisions made during the development of the original source. Different languages have different requirements for space, and different locales have different conventions for layout and formatting. Keeping these differences in mind can help you avoid adding time to an already-tight schedule.

Text expansion

When text is translated, the length of sentences often increases. This occurs for many reasons. Language syntax differs greatly among languages, and what one language can say in five words, another may require eight. Some languages also combine words together to form other words, or they require inflections for each part of speech. Polish, for example, has six cases and three genders (as well as three tenses and imperfect and perfect verb forms). Each requires different case and gender markers for adjectives, nouns, articles, and verbs. The result of this increase is often referred to as *text expansion*.

As a rule, a translated sentence may expand up to 20 or 30 percent in length. What may come as a surprise is that shorter language components (words and phrases) in a target language can expand easily up to 100 percent of the original length of the source. This fact complicates matters when template developers do not account for it in template design, and adjusting layouts for text expansion can add many hours to a localization desktop publishing schedule. Desktop publishing, by the way, can be one of the most expensive aspects of the localization process.

To avoid additional localization costs, design paragraph styles to allow expansion. In Figure 30, the margin for the text doesn't account for expansion in the Russian translation, forcing the Russian side-heading text to wrap.

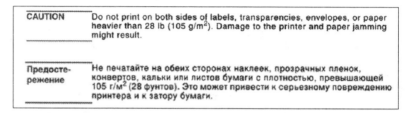

Figure 30: *Text expansion in side headings*

A more localization-friendly approach for this style would be to have the side heading text as run-in text (that is, no hanging indentation) or as a heading above the body of the cautionary text.

Text expansion also causes problems with text in tables, especially if you're designing online help. The problem occurs when short text strings are used in narrow columns. As noted before, a single term or phrase can double in length. In this respect, shorter text strings cause more problems because hyphenation can occur more than once in long words. For some layout applications, columns expand automatically, but in online help and HTML, fixed width table cells have to be resized manually. Depending on how many tables you use, you can add considerable work to the localization schedule.

In Figure 31, some commands take up the entire column width.

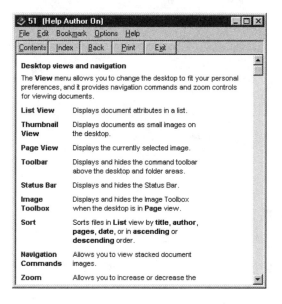

Figure 31: *Table in WinHelp: No room for expansion*

In addition to forcing manual adjustment of the column, such designs also force the window parameters to change to allow the table to expand to the right.

Some table layouts don't cause expansion problems, primarily when they're designed without short text strings in narrow columns. As long as the text has room to expand, the column width is acceptable.

In Figure 32 on page 182, the left column contains icons. Because these do not require room for expansion, the table requires no adjustment.

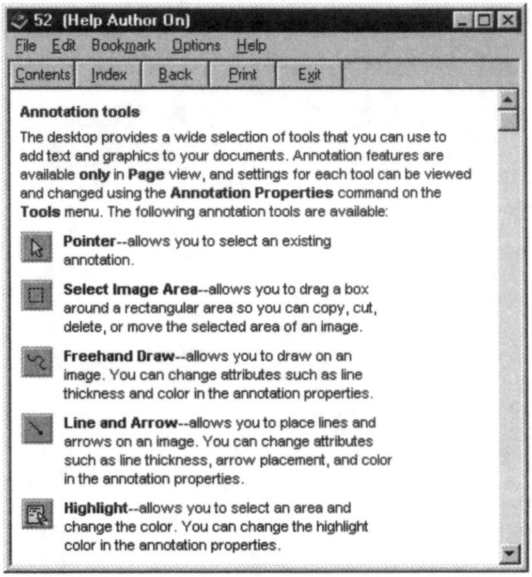

Figure 32: *Table in WinHelp: Acceptable for expansion*

Other problems with tables

A common technique in troubleshooting guides is using if/then scenarios (that is, "If you see this problem, then do this action"). This approach works fine as long as the text occurs in a standard paragraph flow. However, writers often use the same structure to sort lists of procedures. Basing any kind of table on a syntactical construction is likely to cause headaches during localization.

The example in Figure 33 might work well for English, but not all languages use the same word order as English. Some put verbs first or place both subject and object before the verb. How languages combine clauses also varies. During translation, the text in the headings could shift between columns and even between column headings and body cells.

Table 1: Printer Troubleshooting§	
If §	**Then...**§
no lights are on, and the printer won't print,§	check to see that the printer is plugged in and that the switch is in the "on" position.§
the amber light is blink-ing and the printer won't print,§	press and hold the Resume button for two seconds.§
the red light turns on and the printer won't print,§	open the front panel and remove the jammed paper.§

Figure 33: Using syntax to organize tables—a big no-no

In Figure 34 on page 184, the headings have been replaced with nouns that categorize the items in the column. This design works in any language and causes fewer linguistic contortions. Remember that languages vary in syntactical structure and that there is no correlation between word length in one language and word length in another.

Table 2: Printer Troubleshooting§	
Problem§	**Solution§**
No lights are on, and the printer won't print.§	Check to see that the printer is plugged in and that the switch is in the "on" position.§
The amber light is blinking, and the printer won't print.§	Press and hold the Resume button for two seconds.§
The red light turns on, and the printer won't print.§	Open the front panel, and remove the jammed paper.§

Figure 34: *Using nouns for headings—a much better choice*

Writers who are unfamiliar with a text processing tool may resort to poor design practices to create the appearance of a table. For example, instead of using the tool's table feature to design a table, the writer may use tabs and forced breaks to create "columns." Such a design is bound to create problems in localization (Figure 35).

```
PC-Based >        >        >      DW-1500
Alarm Functions   >        >      On-Board Alarms
Can be used for all device types  >   DW-1500 only

Setup can be done on- or offline. >   Setup must be done online.

Individually configurable    >       Fixed polling interval system
polling intervals per alarm  >       based on Onboard Alarm Check
         >        >        >         Interval
```

Figure 35: *Tabs for tables—a localization nightmare waiting to happen*

First, the setting of tabs requires writers to know where boundaries between words can be placed. When text is translated, these boundaries change. Because the tab placement can't adjust automatically, the format of the table is completely destroyed. Second, because of variance in syntax, phrases may have to be relocated in the flow of the table, and what may require a single line in English may require two or three in a target language, so a table may have to be completely re-created to accommodate each language.

Other issues

Some other localization issues to consider are:

- Sort orders—Locales have different conventions for sorting lists, including indexes and glossaries. Any techniques for automatically sorting items will save time. For example, sorting glossaries can be very time consuming. However, by placing glossary entries into cells of a table with no rules (an "invisible table"), localization personnel can often use an application's table sorting capabilities to quickly reorder a translated list. Otherwise, each term has to be moved manually. Also, be aware that some indexing tools and search extensions for online help and HTML do not work for all character sets. The broader your language set is, the more carefully you need to check for language support.

- Cross-reference formats—Word order in cross-references formats is likely to change, and sometimes a single cross-reference in one language has to be made into two cross-references for another to accommodate case or gender differences. One way to avoid this problem is to remove all hard-coded text from the format and only keep the autogenerated portion as part of the format—that is, the heading text or page number. Single-source documentation can make this scenario more difficult, but you do have options. For more information about single sourcing, see Chapter 13, "Single sourcing."

- Autogenerated text—Using autogenerated text (such as a table of contents) can reduce the word count during localization, so use it, but make sure to always edit it at the source. Never edit a table of contents or index by hand; it will only have to be edited again when you regenerate the files. Always edit the headings or the index markers as they occur in the content. Also, make sure to define the styles for the autogenerated text to eliminate formatting by hand.

- Adherence to templates—Following the formatting template for your document can help save a lot of time and expense during localization desktop publishing work. Styles give you global control over the look of paragraphs and characters, and a change that could affect dozens of paragraphs in dozens of files can be changed with a single button click instead of an hour or more of manual reformatting. Avoid using layout overrides (such as changing a single heading paragraph to fall after a page break). Instead, design specific styles for format variations. Use character styles for changes in the dominant font for a paragraph. Style overrides (for example, making a word bold or italic by selecting a button on the toolbar or a choice on the **Font** menu) can create a lot of manual reformatting. The more global you make the style control, the easier your documents will be to localize.

If your department's template defines structure, it's important that your content adhere to that structure. See Chapter 14 for more information about how structured templates enforce consistency in content, which makes the internationalization process smoother.

Think globally, act locally

Even if you know that your company's current documentation strategy doesn't include localization, play it safe and design with an international audience in mind. What may be true for the company's strategy today could change dramatically in a year. Technical and medical businesses are becoming more focused on the world marketplace, and with that shift in focus comes national and cultural pressures to respect the needs of a broader audience. "Thinking globally" from the start will eliminate a lot of rework that you may have to do to make your documentation acceptable in international markets.

A myth still exists in the business world today that English is "the language of international business." Aside from the fact that this "truism" isn't true, it doesn't follow that all cultural references and allusions are clear to someone who has grown up in a different locale. Remember, these people are your audience, too. Write accordingly.

13 Single sourcing

What's in this chapter

- ❖ The traditional workflow
- ❖ Evaluating whether single sourcing is right for a project
- ❖ Benefits of single sourcing
- ❖ Objections to single sourcing
- ❖ Planning for single sourcing
- ❖ Choosing single-sourcing tools

Single sourcing refers to a writing process in which you create multiple deliverables from one set of files.

Applications of single sourcing include:

- Creating two (or more) versions of a deliverable— Suppose the administrator and user guides for a product have similar content, but the administrator book contains programmer-level information that is not appropriate for the user guide. You can use one set of files to produce both manuals instead of creating a set of source files for each book.

Similarly, you can create guides for similar products that share some features.

- Creating multiple output media—If you need both printed and online versions of product information, you can set up files that produce both types of output.

Because single-sourcing processes can be highly automated, it's possible to create different deliverables and output types with little or no manual conversion or post-processing cleanup.

NOTE: In many single-sourcing environments, authors create multiple versions of a document *and* then different output types for the versions; the two are not necessarily exclusive.

Successful single sourcing requires quite a bit of planning. You can, however, use single sourcing create high-quality deliverables *and* save money and time. If you are the lone technical writer in your company or if your department is understaffed, single sourcing could be the only approach that makes it possible to complete multiple deliverables on time and within your budget.

The traditional workflow

Without single sourcing, a documentation department would typically use one of two approaches to complete multiple, related deliverables:

- Developing deliverables simultaneously (parallel development)
- Developing one deliverable first, and then the other (serial development)

Parallel development

Parallel development means that different versions are are created, in separate files, at the same time. In the case of documents with similar information, one writer completes one version of the document while another writer works on the other version. When creating printed books and online help, parallel development often means that one technical writer creates the printed book and another creates the online help.

This approach has a few advantages:

- Both deliverables are ready at the same time.

- Writers can specialize in print or online work and optimize the information for each deliverable, or they can focus on writing for a specific audience (for example, user-level documentation vs. more complex administrator-level material).

But there is also a significant disadvantage. Parallel development is very labor-intensive and maintenance is problematic because you're creating the same information twice. This duplication of effort can also lead to differences in terminology in the multiple versions, which confuses the readers.

Parallel development does make sense when the information in the deliverables is different. However, when there is content overlap, parallel development is time consuming and inefficient.

Serial development

Serial development could be considered single sourcing. You write the information once, and then you strip out or add new information for a similar product, or you convert the source files from one format to another. Serial development has some advantages:

- Because the information is written once, information is consistent across all deliverables.

- The second deliverable is not created until the first deliverable is completed, so the information is finalized.

- Maintenance is simplified because you only convert once per release and do not have to maintain two sets of documentation.

- One writer can first create the first deliverable and then complete the second, so it's less expensive than having two writers working in parallel.

But there are some serious disadvantages:

- Serial development means that one deliverable will lag significantly behind the other. For example, the printed version of a document might be ready three or four weeks before the online help. This leads to scheduling problems before a release because you have to build in several weeks for the conversion and reformatting process. A similar delay occurs when you're creating slightly different versions of a document. The time it takes to strip out nonapplicable material, add new information, and change terminology creates a lag time between deliverables.

- When you're creating printed and online formats, the print tool and the online help tool sometimes do not work well together, so formatting is lost when you transfer information from one tool to the other. The cleanup that's done in the second version must be repeated for each release (unless you keep the files and only put in the changes, in which case you've switched to a parallel development process).

Many companies use serial or parallel development. However, single sourcing provides them with another option. In a single-sourcing workflow, you create a single set of files that contains all the information for multiple versions, different output media, or both.

Evaluating whether single sourcing is right for a project

Single sourcing is not appropriate for every project. If the content of the various deliverables does not overlap, there's little reason to use single-sourcing techniques. For example:

- Unless two products share several features, it's not helpful to combine information about both products into one set of source files.

- If the content of your online help has little in common with your printed user guide, there's no sense in trying to create the help and the printed user guide from one file set.

Benefits of single sourcing

Using one set of files to produce multiple deliverables does offer substantial benefits for your documentation effort, including the following:

- Reducing time to market
- Minimizing errors and inconsistencies

- Saving money
- Presenting information customized for each delivery medium

Reducing time to market

Because multiple deliverables are generated from the same source files, you no longer have to maintain two (or more!) sets of content in a parallel development process.

Working with one set of source files reduces the amount of time necessary to complete the documentation for a product. Less time spent on documentation can help get a product to market more quickly. Also, once a single-sourcing process is set up, it takes less time to update documentation for new releases of a product.

Minimizing errors and inconsistencies

Maintaining the same information in two locations is obviously time consuming (you have to make every update twice), but there's another, less apparent cost. Maintaining multiple information sets increases your chances of making mistakes or adding inconsistent information.

If you have to make 100 changes to your content, you would make 100 changes in a single-sourcing environment. But if you're maintaining two sets of source files, you have to make 200 changes. Or, more accurately, you have to make 100 changes two times. This is extremely tedious for the writer and can lead to additional errors.

Saving money

At a minimum, single sourcing eliminates the problem of making updates in two places. This alone can provide significant savings. But you can also eliminate the costs associated with maintaining multiple file sets. This can make the other savings look puny—especially if your content is later translated into other languages.

You also save money because of increased efficiency and cleaner files.

Presenting information customized for each delivery medium

In addition to saving money and eliminating the need to maintain multiple file sets, single sourcing lets you develop a template for each delivery medium that takes full advantage of that medium. For example, a glossary might be delivered as a series of pop-up topics in online help, but it can be the standard alphabetical list in a book. Reference information can be hyperlinked and made searchable online. Detailed conceptual drawings are probably best left in the printed, high-resolution book. But in online help, hyperlinked flow charts are possible.

Making the business case for single sourcing

Consider the case of a typical software company. The company has approximately 6,000 pages of documentation in a total of 10 books. The material is converted to online help and is translated into eight languages.

The writers develop books in a text processing tool and then save the files to rich text format (RTF), which preserves fonts, italics, and other formatting. The writers import the RTF files into an online help development tool, and then they clean up formatting problems to create WinHelp. Typical formatting tasks include reimporting graphics (which are lost during conversion), recreating hyperlinks and cross-references (lost during conversion), breaking material into topics, creating pop-up topics, and correcting numbering problems. On average, the cleanup process for a 200–300 page book takes about three weeks (120 hours). At a very conservative $30 per hour cost, that works out to $36,000 in conversion costs for 10 books. Therefore, the conversion costs for the English source and all eight languages means a total of $324,000 (9 languages x $36,000)—and that doesn't include translation costs.

The company now introduces a single-sourcing process. The conversion to online help is handled by converting the source files to online formats with an automated mapping tool. The conversion process takes just one hour per book! The company incurs approximately $13,000 in one-time costs for training, template development, software licensing, and consulting. Even with a much higher conversion rate of $120 per hour (versus the $30 for the conversion with the online help development tool), the recurring conversion cost is only $10,800 (1 hour x $120 per hour x 10 books x 9 languages). Translation costs aren't included in this total, either.

The old process costs about $324,000 per release. The new process costs about $23,800 for the first release—and that includes the $13,000 in one-time setup costs.

In short, the cost savings are enormous. If this company produces two releases per year (one every six months), the single-sourcing solution will save more than half a million dollars in the first year!

Objections to single sourcing

The biggest objection to single-sourcing concerns using one set of files to create multiple outputs. Single-sourcing opponents believe that it's impossible to write content that works well both in printed and online documents. They say that the presentation requirements for print and online are too different for a common source file. It's certainly true that some single-sourcing efforts have produced mediocre results. For example, you've probably seen online help that read more like a book—windows full of conceptual information that a user really doesn't want when trying to figure out how to perform one task, and text that didn't take advantage of online hypertext linking.

It's quite possible to write content that works in several output formats. Extensive planning and using the right tools can give you complete control over the content and appearance of multiple deliverables generated from one set of files.

The 90 percent point

Although it's probably true that "hand-crafting" printed and online materials separately will result in higher-quality results, a solid single-sourcing process can produce very good results at a fraction of the cost. This is the 90 percent point. If you can achieve 90 percent of the "hand-crafted" quality at a fraction of the price, you must decide whether the cost you incur to achieve that last 10 percent is worth the effort.

Planning for single sourcing

Planning is an important part of any writing project, but for a single-sourcing project, it's required. There several factors you must consider before applying single-sourcing techniques.

Considerations for creating multiple versions of a document

When setting up files for multiple versions of a document, you need to determine how much information is shared among the deliverables and how much is unique, and what terms and names differ across the deliverables.

Determining what information is shared

The information that is shared across documents provides the backbone for each version.

Suppose you're documenting a printer that has three different models, and you need a user guide for each. You need to map out which features are shared and which are unique. You can create a table to figure this out. Table 5 shows a feature map for different models of a printer.

Table 5: *Mapping the features of three printer models*

Printer model number	500-sheet paper drawer	Secondary paper drawer	Color printing	Duplex printing attachment
Model A-1000	X	X		
Model B-2000	X	X	X	
Model C-3000	X	X	X	X

Based on the table, information about the 500-sheet paper drawer and the secondary paper drawer would go in all manuals. However, information about the duplex printing attachment (which enables printing on both sides of the paper) would go only in the manual for Model C-3000.

Determining similarities and differences in terminology

Even if two or more products have overlap in their features, the products most likely don't have the same name, and similar features may not even be called the same thing in the different products. Also, the titles of the products' manuals are probably not the same (particularly if the product name is part of each title). To handle these kinds of differences, you need to come up with a list of placeholders for names and other terminology. For example, you would need a placeholder called ManualTitle for the printer documentation; you would change the definition of that placeholder based on which book you were creating.

Considerations for creating multiple output types

Because components in a printed document become online help components (or vice versa, depending on what tools you are using), you must consider each element of the book and what it will become in the online version or versions.

Information that looks the same in the book can have different functions in the online help, so it's important to examine not just the appearance of the book, but also its underlying structure. The glossary mentioned in "Presenting information customized for each delivery medium" on page 195 is a good example of this.

What information goes into the deliverables?

The first step is to identify your information types. Scriptorium Publishing generally uses four fairly simple ones (described in detail in "Different types of content" on page 83); other documentation groups may use fewer or more categories. Information types include:

- Interface information, which describes the user interface (the information the user sees on-screen)

- Conceptual information, which describes the how and why of a particular product

- Reference information, which provides information about the syntax of a system (for example, a list of commands and their arguments)

- Procedural information, which explains how to accomplish a particular task

For each information type, you need to identify which deliverables should include that information. For example, conceptual information should be in the print version but not in the online help. Users want specific task information from online help and from manuals, so procedural information would go into both the online help and the hardcopy books.

In addition to various types of information, you also have to consider navigational aids. These include the tables of contents, indexes, alphabetical listings, pagination, headers and footers, and any other information that helps readers find their way through a book or help file. Navigation is generally very different from one medium to another; for example, page numbers are a necessity in a hardcopy book, but they are of little use in online help.

A high-level planning grid

As you break down information into different categories, start looking at the high-level elements in the book and figure out what elements they will become in the help (Table 6).

Table 6: *High-level elements*

Element	Treatment
Table of contents	Generally available in book and help. Help may include more levels to make navigation easier.
Introduction	Usually conceptual; often eliminated from help using conditional text.
Glossary	Usually alphabetical in book; often provided as pop-ups from the words inside the topics in help.
Index	Automatically hyperlinked in the online version.
Context-sensitive interface help	Often, no interface information is provided in the book.
Section	Book sections become topics in the online help.

A paragraph-level planning grid

After assembling a grid for your high-level book elements, you can move on to paragraphs, graphics, tables, and other paragraph-level elements.

In a book, you may have a Body tag that serves more than one function. For example, it might be used for body text but also for the definition of glossary terms. In a single-sourcing environment, you want a one-to-one correspondence between the paragraph styles and their functions, so you would probably need to create a GlossaryDefinition tag. For a print-only book, this is unnecessary, but if you plan to convert the book to online help, you need to be able to identify glossary text uniquely.

Headings generally indicate the start of help topics, so it's a good idea to make sure that the document is broken up into reasonably-sized sections.

In a book, you may have a See Also listing at the end of a section. In the online help, this might become a "Related Topics" list, or just a list of links.

Choosing single-sourcing tools

Choosing the right tool is as essential as planning a single-sourcing process. To ensure your process is as efficient as possible, a tool must provide particular features, as explained in the following sections.

Considerations for creating multiple versions of a document

When evaluating a tool for creating multiple versions of a document, consider how well it lets you complete the following tasks:

- Labeling text for selectively displaying content
- Assembling chunks of text into a document
- Defining placeholders

Labeling text for selectively displaying content

One way to specify what material goes in each version of the document is to label text so that you can selectively display it.

FrameMaker software does this very well with a feature called *conditional text.* You can show or hide text based on the tags you apply to content.

You could use conditional text to show and hide information about the three printer models described in Table 5 on page 198. According to that table, you would not need to apply a condition to the text about the 500-sheet and secondary paper drawers because that information applies to all models. However, you would need to apply a condition to the text about the duplex printing attachment so that it is displayed only in the book about Model C-3000, as shown in Figure 36 on page 204.

Another way to label text for selective display is through attributes that you assign to content. You can think of attributes as hidden labels writers use to organize their content; users of documents don't see them. See "Element attributes" on page 217 for more information about attributes.

Assembling chunks of text into a document

You can also control what information goes into each version of a document through the use of *modules.* Modules are chunks of information that you use to build a document; for example, one chunk for the printer documentation could explain how to attach the duplex printing attachment. When a particular chunk of information does not apply to a printer model, you would not include it in that model's documentation.

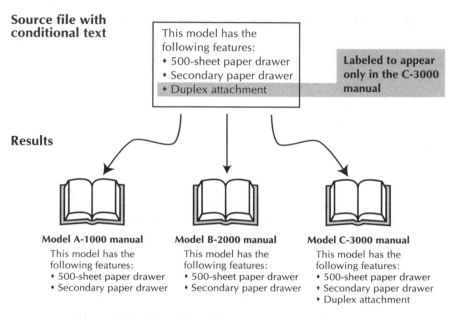

Figure 36: *Conditional text usage*

In addition to giving you the ability to control what information appears in a document, modular documentation also provides the following benefits:

- Easy reuse of content—Reusing and sharing information in modules eliminates rewriting, which wastes time and can introduce inconsistencies in terminology and presentation.

- Automatic updating of embedded modules—When you modify information in a module, it is automatically updated in the documents where it is embedded. Automatic updating eliminates the need for figuring out where modules are used in your source files and making the same changes in the different locations. Making the change in a single location also ensures consistency.

You can find more information about creating modular documentation in *Single Sourcing: Building Modular Documentation* by Kurt Ament (ISBN 0815514913).

Defining placeholders

Placeholders let you easily update product names and other terminology in different versions of a document. Continuing with the example about the three printer models, you could create a Model placeholder in the source files to handle the three model numbers. (You would also need placeholders for any other features and terms that do not have the same name across all models.) You could then create three separate files that contain just the placeholder definitions for each model. Before creating the deliverable for each model, you would import the definitions into the source files (Figure 37 on page 206).

Considerations for creating multiple outputs from one document

When evaluating a tool for creating multiple outputs, a primary consideration is how much it automates the conversion process. For example, some tools require you to reapply some formatting in converted output, but other applications create output that requires minimal or no post-processing.

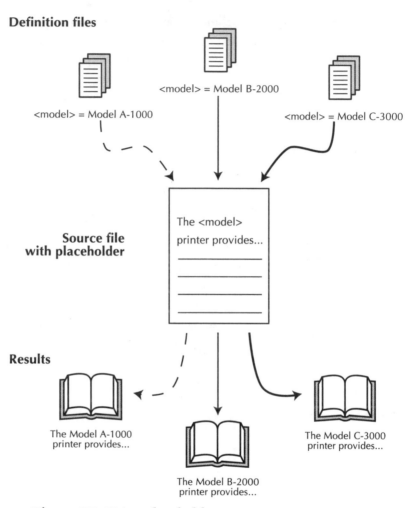

Figure 37: *Using placeholders*

Another factor to consider is how widely implemented the single-sourcing process will be—is it a small group of writers doing the conversions, or is it a solution that is implemented across a large department or company?

NOTE: As an entry-level writer, you probably won't be tasked with evaluating single-sourcing tools. In fact, you'll probably be told what tool to use—and exactly how to use it. However, if you're part of smaller department, you may find yourself in the position of choosing a tool and setting up the single-sourcing process.

Level of automation

Single-sourcing tools vary in how much they automate the process of creating online formats from document source files.

There are tools that automate a good portion of the conversion process but produce output that requires some post-processing clean up. These tools are fairly inexpensive, relatively easy to use, and do not require programming ability. In some cases, you have to save your source files to another format before importing them into the tool—which can remove some formatting and graphics that you have to put back in yourself.

Other tools that provide a higher degree of automation are a better choice. Often, these tools don't require you to reapply formatting or manually save your files to another format before conversion.

Some of these tools require a lot tagging and special formatting in the source files, and you need some low-level programming skills to set up the conversion process. Tools and technologies in this category include the following combinations:

- FrameMaker with MIF2GO
- FrameMaker or Microsoft Word with WebWorks Publisher

- Content based on Extensible Markup Language (XML) transformed into online formats with Extensible Stylesheet Language (XSL)[1]

These advanced tools and technologies provide the ability to set up totally automatic single sourcing. It is possible to build a process that requires writers to do little more than push a button to start conversion. Strict adherence to templates is required, and you may need the help of a consultant to create the templates and to set up the process. Figure 38 shows a single-sourcing workflow you can set up using advanced tools.

Another tool that offers a higher degree of automation is AuthorIT. Unlike WebWorks Publisher and MIF2GO, which convert source files created in another program, writers create the source files in AuthorIT and store them in a database that is part of the application. (The database also provides a way to control who has access to particular content and a way to track versions of a document.)

Using the product's interface, writers assemble a document from chunks of stored content and then specify what type of output they want—Word documents for printed documentation, HTML for web sites, and so on.

---

1. XML is a specification for storing structured information in a text format. XSL is a technology that transforms content based on XML into formats such as HTML and HTML Help. Chapter 14, "Structured authoring with XML—the next big thing," describes structured information, XML, and XSL.

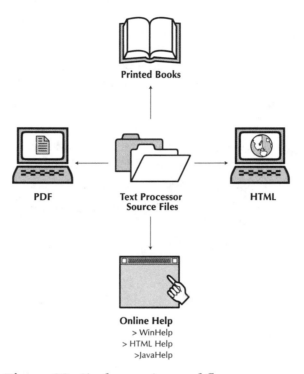

Figure 38: *Single-sourcing workflow*

The conversion process is push-button with AuthorIT, but someone must first set up the formatting standards for the content (just as you do in templates for Web-Works Publisher and MIF2GO, but less scripting is involved). Also, if your department uses AuthorIT, someone must handle the administration of the data-base (determining users' access to content, and so on).

In general, more automation requires higher software costs and a greater degree of complexity in initial setup, but a highly automated process requires writers to spend significantly less time converting files. So, when evaluating the cost of a single-sourcing process, don't focus on just the price of licensing the software. Be sure to also consider the following:

- Any start-up costs, particularly if you need help from a consultant in establishing the process and being trained on how to use it

- The time (and money) it takes authors to convert documents each time there is a documentation release

Level of implementation

Another factor affecting the cost and complexity of single sourcing is the level of implementation. Some smaller departments may handle conversion by using the applications described in the previous section; staffers use applications installed on their computers to create online output. The output is then stored on their machines or on a shared server. The content is distributed to users on CDs or as a help system integrated with the documented application itself. The output may also be stored on the company's web site so users can access it.

Larger companies, however, may set up what's called an enterprise-level solution. The cost of software and implementation can skyrocket to $50,000–$100,000 or more; the software can be highly customized versions of existing single-sourcing applications or even custom-designed applications. Enterprise-level environments usually include significant amounts of networking and database storage capabilities. For example, writers working in different locations generate content that is stored in the same database.

These large-scale solutions are appropriate only for very large organizations, and their implementation requires the services of a consultant—heavy-duty programming and database skills are necessary. These custom-designed environments can provide repositories for information and the ability to output to many media, including on-demand distribution via the Internet. For example, a company's web site can be connected to a database housing the online output. The database displays specific content based on the search terms typed by a user.

14 Structured authoring with XML—the next big thing

What's in this chapter

- ❖ What is structured authoring?
- ❖ What is XML?
- ❖ The impact of structured authoring and XML on writers
- ❖ Does structured authoring work with single sourcing?

With any kind of product or service, there is always something that will be "the next big thing." For example, TV viewers have heard for years about the possibility of watching movies on demand by selecting films from an on-screen menu. That service is now available from many satellite and cable television providers.

Things are not that much different in technical writing—new, more efficient processes are always being developed. You just read about single sourcing in the previous chapter. The latest emerging trend in technical communication is structured authoring with XML. This chapter explains structured authoring and XML, and it describes how they affect the documentation process—and you as a technical writer.

What is structured authoring?

Structured authoring is a publishing workflow that defines and enforces consistent organization of information in documents, whether printed or online. In traditional publishing, a style guide lists content rules, and an editor reviews content to ensure the information conforms to the approved styles.

A few simple examples of content rules are as follows:

- A heading must be followed by an introductory paragraph.
- A bulleted list must contain at least two items.
- A graphic must have a caption.

In structured authoring, these rules are captured in a structure definition document. Writers work in software that *validates* their documents; the software verifies that the documents they create conform to the rules in the structure definition document.

Consider, for example, a simple structured document—
a recipe. A typical recipe requires several components:
a name, a list of ingredients, and instructions. The style
guide for a particular cookbook states that the list of
ingredients should always precede the instructions. In
an unstructured authoring environment, the cookbook·
editor must review the recipes to ensure that the author
has complied with the style guideline. In a structured
environment, the recipe structure *requires* the specified
organization.

NOTE: As a new writer, you generally don't need to
know how to create or modify a structured definition
document, which can be quite complex. Instead, you
just write content based on the rules established by the
document.

Elements and hierarchy

Structured authoring is based on elements. An *element* is
a unit of content; it can contain text or other elements.
You can view the hierarchy of elements inside other elements as a set of nodes and branches.

Elements can be organized in hierarchical trees. In a
recipe, the ingredient list can be broken down into ingredients, which in turn contain items, quantities,
and preparation methods, as shown in Figure 39 on
page 216.

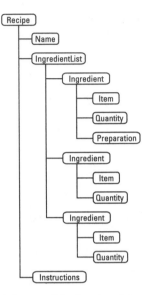

Figure 39: *Recipe hierarchy*

The element hierarchy allows you to associate related information explicitly. The structure specifies that the IngredientList element is a child of the Recipe element. The IngredientList element contains Ingredient elements, and each Ingredient element contains two or three child elements (Item, Quantity, and optionally Preparation).

In an unstructured, formatted recipe, these relationships are implied by how the type looks—for example, the recipe name is in large bold type, and the ingredient list items are in smaller type. The publishing software, however, does not capture the hierarchical relationship between the recipe name and the ingredient list—or the relationships of the components that make up each ingredient list item (item, quantity, and preparation).

In structured documents, the following terms denote hierarchy:

- Tree—The hierarchical order of elements. (By the way, it's not unusual to think of a family tree when considering the hierarchy of elements.)

- Branch—A section of the hierarchical tree.

- Leaf—An element with no descendant elements. Name, for example, is a leaf element in Figure 39.

- Parent/child—A child element is one level lower in the hierarchy than its parent. In Figure 39, Name, IngredientList, and Instructions are all children of Recipe. Conversely, Recipe is the parent of Name, IngredientList, and Instructions.

- Sibling—Elements are siblings when they are at the same level in the hierarchy and have the same parent element. Item, Quantity, and Preparation are siblings.

Element attributes

You can store additional information about the elements in attributes. An *attribute* is a name-value pair that is associated with a particular element. In the recipe example, attributes might be used in the top-level Recipe element to provide additional information about the recipe, such as the author and cuisine type (Figure 40).

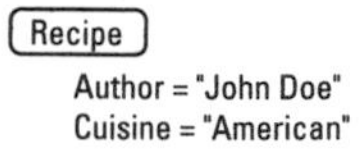

> Recipe
> Author = "John Doe"
> Cuisine = "American"

Figure 40: *Attributes capture additional information about an element*

Attributes provide a way of further classifying information. If each recipe has a cuisine assigned, you could easily locate all Greek recipes by searching for the attribute. Without attributes, this information would not be available in the document. To sort recipes by cuisine in an unstructured document, a culinary expert would need to read each recipe.

When structured content is stored in a database, the attributes provide a way to search for specific information—think of attributes as information about information (also known as *metadata*).

Formatting structured documents

While writing a structured document, you don't apply paragraph formats to text as you do in unstructured word processing programs. When you put text in an element, formatting is assigned based on the element. Some applications immediately apply formatting that you can see in the application's interface; other applications apply the formatting later when the content is processed for output.

What is XML?

Extensible Markup Language (XML) defines a standard for storing structured content in text files. The standard is maintained by the World Wide Web Consortium (W3C).[1]

---

1. Detailed information: http://www.w3.org/XML/

XML is related to other markup languages, including:

- Hypertext Markup Language (HTML), which you may have used to build web pages. HTML is one of the simpler markup languages; it has a very small set of tags.

- Standard Generalized Markup Language (SGML), which is a heavy-duty markup language. Implementing SGML is an enormous undertaking. Because of this, SGML's acceptance has been limited to industries producing large volumes of highly structured information (for example, aerospace, telecommunications, and government).

XML is a simplified form of SGML that's designed to be easier to implement.[1] Publishing applications require complex frameworks to represent the structures of technical documents—including parts catalogs, training manuals, reports, and user guides—and XML can provide those structures.

XML syntax

XML is a markup language, which means that content is enclosed by tags. In XML, element tags are enclosed in angle brackets:

```
<element>This is element text.</element>
```

A closing tag is indicated by a forward slash in front of the element name.

---

1. SGML vs. XML details: http://www.w3.org/TR/NOTE-sgml-xml-971215

Attributes are stored inside the element tags:

```
<element my_attribute="my_value">This is element
text.</element>
```

Unlike HTML, XML does not provide a set of predefined tags. Instead, you define your own tags and the relationships among the tags. This makes it possible to define and implement a content structure that matches the requirements of your information. Figure 41 shows an XML file that contains a recipe. Compare it to the recipe hierarchy shown in Figure 39 on page 216.

How are XML and structured authoring related?

Structured authoring is a concept. XML is a technology (or, more precisely, a specification) that lets you implement structured authoring using plain text files. The terms XML and structured authoring are often used almost interchangeably.

XML is said to be *well-formed* when basic tagging rules are followed. For example:

- All opening elements have a corresponding closing element, and empty elements use a terminating slash:

```
<element>This element has content</element>
<empty_element />
```

- Attribute information is enclosed in double quotes:

```
<element attribute="name">This is a legal
attribute</element>
<element attribute=name>This is not well-
formed.</element>
```

- Tags are nested and do not "cross over" each other:

```
<element>This is <strong>correct.</strong>
</element>
<element>This is <strong>not correct.</element>
</strong>
```

```
<Recipe Cuisine = "Italian" Author = "Unknown">
    <Name>Marinara Sauce</Name>
    <IngredientList>
        <Ingredient>
            <Quantity>2 tbsp.</Quantity>
            <Item>olive oil</Item>
        </Ingredient>
        <Ingredient>
            <Quantity>2 cloves</Quantity>
            <Item>garlic</Item>
            <Preparation>minced</Preparation>
        </Ingredient>
        <Ingredient>
            <Quantity>1/2 tsp.</Quantity>
            <Item>hot red pepper</Item>
        </Ingredient>
        <Ingredient>
            <Quantity>28 oz.</Quantity>
            <Item>canned tomatoes, preferably San Marzano</Item>
        </Ingredient>
        <Ingredient>
            <Quantity>2 tbsp.</Quantity>
            <Item>parsley</Item>
            <Preparation>chopped</Preparation>
        </Ingredient>
    </IngredientList>
    <Instructions>
        <Para>Heat olive oil in a large saucepan on medium. Add
        garlic and hot red pepper and sweat until fragrant. Add
        tomatoes, breaking up into smaller pieces. Simmer on
        medium-low heat for at least 20 minutes. Add parsley,
        simmer for another five minutes. Serve over long pasta.
        </Para>
    </Instructions>
</Recipe>
```

***Figure 41:** A recipe in XML*

XML is said to be *valid* when the structure of the XML matches the structure specified in the structure definition. When the structure does not match, the XML file is *invalid* (Figure 42).

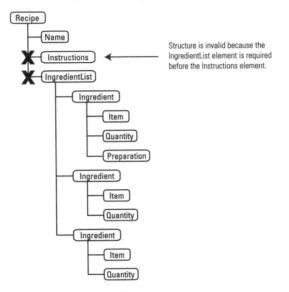

Figure 42: Invalid structure

Defining structure in XML

Just as structured authoring relies upon a document that defines the hierarchy of elements, XML documents are based upon a definitions file. For example, a Recipe element definition might read as follows:

```
<!ELEMENT Recipe (Name, History?, IngredientList,
Instructions)>
```

The preceding definition states that Recipe element contains a required Name element, an optional History element, and required IngredientList and Instructions elements.

Unlike the definitions file for a structured authoring tool, the file for XML documents does not contain any formatting information because XML is text based. However, when you open XML content in some authoring tools, the tool recognizes all the elements *and* knows what formatting to apply because the authoring tool assigns formatting based on element names.

The impact of structured authoring and XML on writers

XML and structured authoring result in a completely different way of looking at information. Instead of following the familiar page- and paragraph-based method, structured authoring requires that writers consider information as a hierarchy with a separate formatting layer (Figure 43 on page 224).

In a structured workflow, writers still create content, but they have fewer formatting and publishing responsibilities—these tasks are generally automated by the structure definition. You assign an element to text, and formatting is automatically applied based on the element's definition. You no longer need to worry about applying paragraph tags or knowing which tags are allowed in particular contexts.

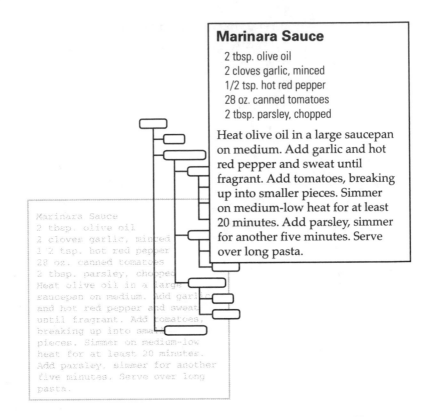

Marinara Sauce

2 tbsp. olive oil
2 cloves garlic, minced
1/2 tsp. hot red pepper
28 oz. canned tomatoes
2 tbsp. parsley, chopped

Heat olive oil in a large saucepan on medium. Add garlic and hot red pepper and sweat until fragrant. Add tomatoes, breaking up into smaller pieces. Simmer on medium-low heat for at least 20 minutes. Add parsley, simmer for another five minutes. Serve over long pasta.

Figure 43: *Representing a document as a series of layers*

Instead of wrestling with content rules and formatting problems, you focus on content and organization—typically a better fit for a writer's skills and interests than desktop publishing. Working within a structure increases your productivity and improves the quality and consistency of the final output.

NOTE: You probably won't work directly in XML files as a new writer. Most likely, you'll use a structured authoring tool that shows you the structure of content. As you get more comfortable with structure and working with tagging languages, you may edit XML content directly in a text editor.

Does structured authoring work with single sourcing?

Single sourcing and structured authoring often go hand-in-hand. Many of the higher-end single-sourcing tools that completely automate conversion work just as well with element-based content as unstructured content.

The consistency created by structure rules is a big plus when it comes to single sourcing. To get the cleanest conversion possible during single sourcing, there can be no overrides to formatting in document files. Validated structured content can minimize and even eliminate such overrides because authoring tools enforce the defined structure and automatically apply formatting.

Also, you can transform XML content into online formats without the use of a third-party conversion tool. An emerging technology, Extensible Stylesheet Language (XSL), enables the conversion of XML content into HTML, HTML Help, and other formats. Setting up the XSL scripts for the conversion process is not for the timid. However, once the process is set up, conversion is highly automated, and there is no additional software cost. XSL consists of free, open-source (but not necessarily simple) code, and there are many freeware XSL processors.

A Getting your first job as a technical writer

What's in this appendix

- ❖ Demonstrating the skills of a technical writer
- ❖ Interviewing
- ❖ The portfolio
- ❖ Where should I look for a tech writing job?
- ❖ Professional organizations
- ❖ Working as a freelance technical writer

After reading all about how the technical documentation process works, you're probably wondering, "But how do I get my first technical writing job?"

Before you start your job search, consider your mission. To get a job as a technical writer, you need to convince an employer that you have mastered the basic skills that every technical writer needs. These are:

- An understanding of technology
- Writing ability
- Organizational skills
- Detective and people skills

Refer to Chapter 1 for a detailed discussion of each of these skill sets.

If you can prove that you have these skills, you can get a job as a technical writer. This chapter will show you how to reassess your current skills and make them relevant to technical writing.

Demonstrating the skills of a technical writer

Although new writers do grapple with "I need experience to get a job, but I need a job to get experience," you can minimize this problem. Experiences as a student, as a worker in another line of work, and as a volunteer can help show a prospective employer you've got the technological, writing, organizational, detective, and people skills required to succeed as a technical writer. You need to tailor your resume to highlight these skills.

Whenever you make changes to your resume, carefully check it for spelling, grammar, and formatting errors—and have a friend with writing skills look it over, too. Errors in resumes submitted for technical writing jobs will often automatically disqualify you from consideration.[1]

---

1. We once received a cover letter and resume that used different spellings of the candidate's first name.

Alterations required

There's a big difference between tailoring your resume for a particular position and lying about your experience. Do not—EVER—lie on your resume. If you get caught, your career will take a hit. But changing your description of a previous job to highlight work experience that's relevant to the job you want is perfectly legitimate.

Another tricky area is software experience. If a particular job requires that you know a certain software package, do not just add that package to your resume. Again, getting caught could have serious consequences. Instead, find a copy of the software and spend some time learning it. Then, put it on your resume. If the employer asks, explain that you are self-taught. In many cases, the fact that you made the effort to learn the software will cancel out your lack of practical experience with it.

Assume that you do add a bunch of software to your resume and get hired on the strength of these fabrications. On day one, you show up on the job and are handed a new project and told to work in one of the software applications that you don't know. Now, you're stuck. You can't request training because you're supposed to know the application. You can struggle without assistance, but then the assignment will take much longer than it would take an expert software user. Either way, your inexperience will become apparent to your coworkers (and your boss). This is not a good way to start off in a new job.

Understanding of technology

You can demonstrate your understanding of technology in many different ways. Perhaps you studied computer science, math, or engineering in college. If so, your education will take care of the "Is this candidate technical enough?" question. If you are one of the many technical writers with a liberal arts degree, you'll need to show that you have technical aptitude. Find a way to prove that you have some grasp of technology and that you are open to learning about new technology when you're looking for a tech writing job.

These days, almost everyone has had some exposure to basic computer software, such as word processors, email applications, web browsers, and spreadsheet programs. Basic computer literacy is a requirement for many jobs, not just technical writing. You'll want to be sure that this computer literacy is apparent from your resume. If you apply for a job by sending a resume via email, your actions prove that you can handle basic tasks on a computer.

Make a list of all the software that you've used. You may be surprised at how many different programs you already know. Add this list to your resume. If you have expert-level knowledge of any packages, be sure to point that out. And don't forget to list the operating systems you've used—Windows, Macintosh, UNIX, and Linux, for example. If you're comfortable working with shell programs, such as csh or telnet, you should specify that as well. These details give the employer reading your resume a good feel for your level of technical knowledge.

If your list of software is painfully short, and if you're none too comfortable using a computer, you may have some work to do before you apply for your first technical writing job. Consider some of the following options.

Public libraries

Most public libraries have computers and Internet access. You can get free email accounts from many different providers—the librarian can probably help you out with this. See "Free email providers" on page 262 for more information.

If you can get computer access at the library, spend some time getting familiar surfing the Internet and using email.

College computer lab

Most universities and colleges have computer labs. If you have access to one of these labs, you should find high-speed Internet access and a suite of word processing and other basic software applications.

Job training

If you're eligible for job training, you may be able to get into computer classes. This could include government programs and outplacement services. Outplacement is most often provided by a company that is laying off a large number of people. If you get caught in a layoff, find out what services your former employer provides and take advantage of them.

Buying a computer

This is the most expensive option, but it might also be the best one. Having a computer in your home means that you can practice your computer skills any time you want.

Building web pages

Many people build personal web pages. This is a great way to put technology to work for you and to show that you understand Internet technology. If you have some experience with web pages, list that on your resume.

If you've built web pages or a web site, consider providing the web addresses (URLs) of some web pages that you've built. But be careful—if your pages are full of strong personal opinions or political information, showing them to a potential employer could backfire in several ways. Political pages could cost you a job offer because the person reviewing your resume doesn't agree with your politics. Similarly, pointing to the web site for a particular church or other religious organization is asking for trouble. But if you helped put together a web site for your local animal shelter, it can't hurt to list that web site on your resume.[1]

If you were employed by an organization with a strong political viewpoint but didn't really agree with their politics, you might consider putting a disclaimer on your resume: "Built web pages for *xxx* organization. Do not necessarily endorse their point of view."

If you built the pages from scratch by writing HTML code in a text editor, say so. If you used a web design tool, specify which tool or tools you used. Don't forget about volunteer work, particularly if you're helping maintain the web pages (or even the web server) for a nonprofit group.

Cheap software

Many software manufacturers offer trial versions of their software that you can use to teach yourself. For example, Adobe (www.adobe.com) offers a trial version of FrameMaker that does everything the licensed version does, except you can't save or print files. Trial software is sometimes bundled with third-party books about applications. Tinkering with a trial version is an excellent way to learn a program. Also, if you're a student, check with software companies about buying academic versions of software. Academic versions are

---

1. We've received resumes pointing to pages with topics including feminist lesbians in China (no, really), fathers' rights, really terrible poetry, and much more.

usually the same as full-license versions, but they cost significantly less—basically, you get a student discount, and it can be quite hefty.

If neither of these avenues works, try some of the online auction sites, such as ebay.com. You may be able to buy software below market price. Be careful, though, to check the seller's credentials; some of the software sold on auction sites is not legitimate. If you buy pirated software, you won't be able to upgrade it.

Software at work

If you're already in another line of work, there's a chance you may be using software specifically for your field. Even though the tool may not be used by technical writers, you should still list it on your resume to show that you use the particular industry's tools. You need to demonstrate you can master the tools of any trade.

Writing ability

Students, particularly liberal arts majors, can demonstrate writing ability by using a good term paper as a writing sample. Even if the paper is not about a technical topic, it can demonstrate that you understand the principles of grammar and how to organize content. Try to find an assignment that demonstrates your ability to perform research and organize the information you uncover into a coherent paper.

Many aspiring technical writers work for campus newspapers or magazines. Highlight that type of experience (and be sure to point out if you had a management position or helped with the design and layout in addition to writing).

Working for a college paper or magazine can also provide excellent writing samples. If you can show both article clippings and academic writing in a portfolio, you prove experience with different types of writing, which is a plus.

Liberal arts degrees *can* be useful

Years ago, many employers considered liberal arts degrees to be as useful as a parka in the Sahara. Today, however, many companies—particularly those who employ technical writers—recognize the value of liberal arts degrees. Because liberal arts curricula generally include a great deal of writing, liberal arts graduates often a gain a better understanding of the craft of writing. Term papers also require research and analysis of a subject (whether Greek history or Shakespeare), and these abilities are very important in technical writing.

There's another less obvious benefit of a liberal arts background. Many liberal arts programs require students to take a wide variety of classes, which forces students to focus on different material every semester. The ability to juggle those different subjects and to switch gears mentally is a very valuable skill for workers in any field. It's particularly important in technical writing, where it's not uncommon to write documentation for a half-dozen different products in a year (and sometimes, you may find yourself writing about two completely unrelated products simultaneously).

If your college days are in the distant past and you're already in the work force, you probably have some writing samples as well. Look for proposals, reports, press releases, marketing brochures, or white papers that demonstrate writing ability. Remember to check into the volunteer work you've done for more possibilities.

See "The portfolio" on page 239 for more information about creating a portfolio.

Organizational skills

Your resume should clearly demonstrate that you can plan a schedule and deal with multiple tasks. Students should list extracurricular activities that involved any sort of managing (whether it be a project, people, or both), and those already in the workforce need to be sure their resume reflects any project management experience. Even volunteer work as a fund-raising coordinator is worth noting, particularly if you handled many different tasks.

Be wary, though, of listing activities that may tell your potential employer more than you want. For example, experience as the social chair of your fraternity or sorority is probably not going to convince your potential employer of your leadership abilities (your partying abilities, maybe, but that's usually not a plus).

Detective and people skills

College students can easily demonstrate detective or analytical skills with academic papers, such as theses, research projects, and term papers. Those already working can point to job responsibilities that require analytical ability. These might include projects where you were asked to evaluate several different options and write a memorandum recommending a particular decision. White papers are excellent examples of writing that requires analysis; if you have written any white papers, be sure to include the best ones in your portfolio and to list the experience on your resume.

If you have had a leadership role in extracurricular activities, you can use this to demonstrate your people skills. But by far the most important test of your people skills will be in an interview.

Interviewing

The personal interview is a critical part of the job hunting process. There are many excellent books on how to interview, but here are some basic pointers.

Dress professionally

The fact that a company has a casual environment does not mean you are not required to dress professionally for the interview. If the company is very formal (typically banks, financial services companies, and government contractors), plan on formal business attire—suit and tie for men, business suit for women. If the company has a casual environment, you can probably dress less formally—dress shirt, pants, and a tie for men; separates for women.

Consider regional differences. For example, Austin is notorious (or maybe noted) for its casual dress codes. "Austin casual" means shorts, a T-shirt, and sandals. Other cities, such as Atlanta and Dallas, tend to be more corporate and have more formal dress codes.

Your goal should be to dress a notch or two better than the employees in the office. Your attire should show the employees that you are capable of looking like a professional. You may not be required to wear pantyhose or a tie for the next three years if you get the job, but it's a good idea to prove that you know how.

Don't oversell

Many job-hunting books tell you to sell yourself in the interview. You do need to do that, but don't overdo it. In addition to convincing the employer that you're right for the job, the interview is also an opportunity for you to assess whether the job is right for you. Ask questions about how the company works, what your responsibilities would be, and evaluate whether those responsibilities are what you want.

The interview is about figuring out whether there's a fit between you and the job. If you respond enthusiastically to every question ("I just *love* taking notes in meetings!"), you risk leaving the impression that you would be happy to perform duties that perhaps you don't really want to do.

Don't ask about salary

Asking about salary puts you in a very poor negotiating position. Wait until the employer brings up the topic. Many companies will ask you what salary you're looking for. The best response to this is to inquire about the salary range for the position. If they give you a range, you can say, "That sounds reasonable. I'm sure we can come to an agreement."

If you have previous relevant experience, you may be able to negotiate an offer higher than the base salary for an entry-level technical writer with no experience.

Starting salaries for technical writers vary tremendously by region. Take a look at the STC web site (www.stc.org) for a salary survey. The U.S. Department of Labor also provides statistics on average salaries.

Be on time

Be on time for your interview. If you're not familiar with the company or the area, scout out the company the day before to ensure you know how to find it. Plan for traffic. If you arrive 15 minutes early, you can always sit in the parking lot for a few minutes before you go in. You should arrive at the reception desk (or at the interviewer's office) a little early, by five minutes or so. Do not show up earlier and expect the interviewer to see you immediately.

If you're going to be late, call and let the interviewer know immediately. Apologize, tell him or her how late you're going to be, and offer to reschedule for another time. If the problem is due to circumstances beyond your control (your plane has been delayed), point that out. If you're going to be late because the baby threw a fit when you dropped her off at daycare, you might want to be nonspecific about the problem and leave home earlier next time.

Send a thank-you note

Sending thank-you notes for interviews has gone out of fashion, and many applicants don't bother. You should. Sending a note immediately after the interview reiterating your interest in the position will make you stand out from all the other candidates who didn't bother. It also gives you an opportunity to mention something you might have forgotten to say in the interview. You can can send your note via email, but sending it by mail makes a better impression.

What federal laws prohibit employment discrimination?
Title VII of the Civil Rights Act of 1964 (Title VII) prohibits employment discrimination based on race, color, religion, sex, or national origin.
The Equal Pay Act of 1963 (EPA) protects men and women who perform substantially equal work in the same establishment from sex-based wage discrimination.
The Age Discrimination in Employment Act of 1967 (ADEA) protects individuals who are 40 years of age or older.
Title I of the Americans with Disabilities Act of 1990 (ADA) prohibits employment discrimination against qualified individuals with disabilities in the private sector and in state and local governments.
Section 501 of the Rehabilitation Act of 1973 prohibits discrimination against qualified individuals with disabilities who work in the federal government.
The Civil Rights Act of 1991 provides monetary damages in cases of intentional employment discrimination.
Source: Equal Employment Opportunity Commission web site, www.eeoc.gov

The portfolio

A good collection of writing samples can demonstrate that you have writing and analytical abilities. If you've done different types of writing—term papers, news stories, and business writing (proposals, reports, and so on)—be sure your portfolio reflects the breadth of your writing experience.

However, most people applying for their first technical writing jobs probably don't have any true technical writing samples to show. Remedy this situation by writing some brief passages that show you can write task-oriented material.

You can do something as simple as the cake-baking procedure shown in Figure 5 on page 93, or you could document a few tasks for a piece of software (for example, how to create, save, and edit a file with the Notepad text editor, which comes with the Windows operating system). If you don't have the software (or ability) to draw figures, include placeholders that explain what's in the figures you would add to the procedure.

Be sure to keep a particular audience in mind while writing your tasks, and you might even want to specify the audience on the printout of the sample task you include in your portfolio. You'll really impress an interviewer if you demonstrate you know the importance of writing for your audience—the cardinal rule of good technical writing.

The sample procedures don't have to demonstrate something highly technical; it's more important that they show you can write for a specific audience and know how to break a task down into steps.

What should my portfolio look like?

Professionals such as graphic designers and architects often carry enormous leather folders in which they showcase their best work.

For your samples, you can probably use something a little smaller—writing samples usually aren't oversized. Consider a basic leather folder. You can often find nice ones at your college bookstore or office supply stores. Any professional-looking folder that lets you organize your pieces will work.

Should I bring my portfolio to the interview?

Yes. Always bring your portfolio to the interview, even if the employer does not request it.

If you're interviewing by phone, and the employer wants you to send writing samples, be sure to send copies—it's best to keep your originals. You can also provide the addresses of web sites that show samples of your writing. Do not send any materials that are proprietary or confidential without getting written permission from the company for whom you created them. Sending confidential materials shows that you do not value a company or client's confidentiality requirements, and that will not score points with the interviewer.

Where should I look for a tech writing job?

Most technical writers (approximately 85 percent) work in the high-tech industry. With this in mind, you might start by investigating which cities have strong high-tech industries. Some of the best-known areas include:

- Silicon Valley (San Francisco Bay area), California
- Austin, Texas
- Boston, Massachusetts
- Research Triangle Park, North Carolina
- Portland, Oregon
- Seattle, Washington
- Washington, DC

If your home town isn't on the list, don't despair. Many other cities have significant high-tech concentrations in specific industries, such as:

- Dallas, Texas (telecommunications)
- Wichita, Kansas (aircraft)
- Melbourne, Florida (aerospace)

Contracting—getting your foot in the door

Many beginning technical writers start out working for a contract shop. It's often easier to get a job as a contractor than as a staff employee, especially if you have limited experience. The contract shop has an agreement with the client company. You do work for the client company, but your paycheck comes from the contract shop.

Some things to consider about contracting include:

- Contracting gives you the opportunity to get started and prove yourself. If you do a good job, there is sometimes the possibility of getting hired permanently by the client company.

- As a contractor, you are generally paid by the hour, which means that you'd get paid for any overtime. Staff technical writers are generally not paid overtime.

- As a contractor, you are usually an hourly employee, so you get paid only for the hours that you work. If you get sick, you may not have paid sick leave.

- If the workload decreases, contractors are usually let go before employees.

- A typical entry-level contract usually lasts six months or so. You can then move on to other contracts and gain a lot of experience in just a few years.

Contracting is not ideal, but it is easier to find contract employment than permanent, so it can be a good starting point.

Transferring within a company

If you are already employed but want to change jobs and become a technical writer, consider applying for a transfer. In many large companies, employees get preference over applicants from outside the company. If you can get your manager's approval, a transfer might be just the thing you need to get started as a technical writer.

The end run into technical writing

Many technical writers start out as desktop publishers or editorial assistants and then move into technical writing jobs. If you have the skills needed for a job as a production specialist or an editorial assistant, consider taking that job and then moving into writing after gaining some experience on the tools and editing side of technical communications. But be careful—you probably don't want to stay at the assistant level for more than a couple of years.

Start-up companies

Small companies that are just getting started are sometimes willing to hire new technical writers. The trade-off is that they're generally not willing to pay very much.

But if you can afford it, working for a start-up company could get you that all-important first job—and who knows, you might like the insanity that comes with an entrepreneurial company.

If you've got it, flaunt it

If you've been in the workforce for a while, you probably have some specialized knowledge. For example, if you've worked as a cable technician, you know quite a bit about installing and splicing cable. Try to find a company in which your experience will be relevant and useful. That company could be more willing to hire you as an entry-level technical writer because you already understand their technology.

Internships

Documentation departments sometimes offer summer internships to college students. Getting such an internship can be very valuable because it lets you see how a documentation department really works, even if interns are sometimes delegated less than glamorous work (such as transferring five sets of review comments into a single document or making copies of release notes).

In addition to giving you the opportunity to learn real-world technical writing skills and to observe a doc group in action, an internship can also give you a foot in the door of a prospective employer. It's not unusual for a company to groom interns for future full-time employment (but don't assume that an internship is a guarantee of future employment).

Professional organizations

Professional organizations can help you with your job search by giving you a chance to meet people in the industry and to attend educational seminars. Various email lists are also available. For a detailed list of resources, see Appendix B.

Working as a freelance technical writer

Some technical writers choose to work independently instead of working for a company or as a contractor through an agency. But freelancing is a difficult path, and you should be aware of some of the pitfalls before you try it.

What you need to make it as a freelancer

As an employee, you are provided with many things that you might take for granted: the equipment to do your job, a space in which to work, and sometimes even training. To be successful as a freelancer, you need outstanding skills—both job skills (writing, research, and the like) and business skills (bookkeeping, management, and the basics of banking and cash flow management).

The second major requirement for a successful freelancer is an excellent professional network. Networking and referrals are how you find jobs, so it's critical that you have a wonderful network that works for you. Ideally, you want to get a phone call every week that starts, "So-and-so recommended that I call you about doing some writing for our project." Building this network takes years, so if you plan to become a freelancer some day, now is a good time to start working on your network.

The third talent that you need is the ability to sell. Because technical writers tend to be somewhat introverted, the idea of selling is often horrifying. "Can't I just do a good job?" they wail. Doing a good job is, of course, critical to your freelancing success, but doing a good job doesn't happen until you get the job—and for that, you need the ability to sell. Selling doesn't have to mean acting like a smarmy salesman, but it does mean that you need the confidence to go to a potential client, present a proposal, ask for a particular amount of money, and ask for the project. *If you cannot or will not do any selling, you cannot make it as a freelancer.*

How much experience do I need to begin freelancing?

There are no official rules for this, but it usually takes a new writer about three years to learn the craft. In addition to mastering technical writing skills, you also need business skills. Many writers spend the first year or two of their career working as a contractor, then work as a staff employee for a few years, and then look at freelancing after that. A minimum of three to five years experience is probably a good idea before you dive into freelance work.

We've already given you two mantras:

- The real estate agent's mantra: location, location, location

- The technical writer's mantra: audience, audience, audience

Time for one more. The freelancer's mantra should be:

Network, network, network

B Resources

What's in this appendix

- ❖ General technical writing
- ❖ Editorial
- ❖ Audience and task analysis
- ❖ Information design
- ❖ Indexing
- ❖ Management
- ❖ Single sourcing
- ❖ FrameMaker
- ❖ Other tools and technologies
- ❖ Professional organizations
- ❖ Ergonomics
- ❖ Job banks

This appendix lists some print and online resources for technical writers.

NOTE: The web site addresses in this appendix and in Appendix C were accurate at the time of printing. If an address does not work, you may be able to find the information by going to the site's home page and doing a search on the topic.

General technical writing

Print

- *International Technical Communication: How to Export Information About High Technology,* Nancy Hoft, 1995, ISBN 0471037435. A guide to writing for international audiences.

Mailing lists

- techwr-l

  Mailing list for technical writers, technical editors, and others working in technical communications. Go to the techwr-l web site to subscribe:

  www.raycomm.com/techwhirl/index.php3

  Traffic and off-topic chatter on the list is very high, but you can find some valuable information every now and then. You can also search the archives for particular information.

- stccicsig-l

  Mailing list for technical writers interested in consulting and independent contracting (CIC).

  To subscribe, go to:

  lists.stc.org/cgi-bin/lyris.pl?enter=stccicsig-l

Editorial

Print

- *The Chicago Manual of Style,* The University of Chicago Press, 15th edition, 2003, ISBN 0226104036. The technical editor's bible. The companion web site to this must-have reference is at:

  www.press.uchicago.edu/Misc/Chicago/cmosfaq

- *The Elements of Style,* William Strunk and E.B. White, 4th edition, 1999, ISBN 020530902X. A classic style reference.

NOTE: Your company may distribute its own style guide and dictionary. Be sure you have the latest releases of those documents.

Web sites

- www.scriptorium.com/Standards

  Scriptorium Publishing's style guide.

- www.microsoft.com/downloads/ details.aspx?FamilyID = b494d46b-073f-46b0-b12f-39c8e870517a&DisplayLang = en

  The Microsoft Manual of Style for Technical Publications, Microsoft Corporation. Once available in print, this book is now available as a free download. You may not agree with some of the terminology, but the manual is a valuable reference for computer-specific style issues.

- www.whatis.com

  A site that defines acronyms and technical terms.

Audience and task analysis

- *User and Task Analysis for Interface Design,* JoAnn T. Hackos and Janice C. Redish, 1998, ISBN 0471178314.

Information design

- *The Visual Display of Quantitative Information,* Edward R. Tufte, 2nd edition, 2001, ISBN 0961392142.

- *Dynamics in Document Design: Creating Text for Readers,* Karen A. Schriver, 1996, ISBN 0471306363.

Indexing

- *Indexing Books,* Nancy C. Mulvany, 1994, ISBN 0226550141.

Management

Print

- *Managing Your Documentation Projects,* JoAnn T. Hackos, 1994, ISBN 0471590991. The best-known book about managing documentation projects. Some of the methods covered won't work well at smaller companies, but this book is still a good resource for learning about project management.

Mailing list

- stcmgmtpic-l

  Mailing list for technical writers interested in management issues.

  To subscribe, go to:

  lists.stc.org/cgi-bin/lyris.pl?enter=stcmgmtpic-l

Single sourcing

- *Single Sourcing: Building Modular Documentation,* Kurt Ament, 2002, ISBN 0815514913.

FrameMaker

Print

- *Publishing Fundamentals: FrameMaker 7,* Sarah O'Keefe and Sheila Loring, 2006, ISBN 0970473338.

- *FrameMaker 7 Workbook Series*, Scriptorium Press. Available at www.scriptorium.com/books.

- *Adobe FrameMaker 7.0 Classroom in a Book,* Adobe Creative Team, 2002, ISBN 0321131681.

Web sites

- www.adobe.com

  Adobe Systems makes FrameMaker software. The web site contains product information, support information, printer drivers, and more.

- www.scriptorium.com

  Scriptorium Publishing offers FrameMaker template design, training, and consulting. The web site contains useful white papers and a collection of links to other FrameMaker resources.

- www.frameusers.com

  An unofficial FrameMaker resource site. It contains information about FrameMaker service providers, add-on software, and more. If you're looking for FrameMaker plug-ins, trainers, utilities, and the like, you can't go wrong with this site. It's somewhat graphics-intensive, but it's worth it.

- www.frame-user.de

  Another unofficial FrameMaker resource site. If you speak German, this is the site for you. Even if you don't speak any German, check it out. The web site creator designed the site to look like a FrameMaker document; it's quite a unique look for a web site!

Mailing lists

- framers@frameusers.com

  A mailing list for FrameMaker users. The framers list has been around for at least seven years; its membership includes both new FrameMaker users and experts, some of whom have subscribed to the list for many years. List traffic is heavy but generally fairly well-focused on issues related to FrameMaker. To subscribe, send email to:

  majordomo@frameusers.com

In the body of the message, type:

subscribe framers

- framers@omsys.com

  Similar to the framers list based at frameusers.com. Many topics are posted to both lists. This framers@omsys.com tends to be more focused on power user issues than the other list. The framers@omsys.com list also has less traffic than framers@frameusers.com. To subscribe, send email to:

  majordomo@omsys.com

  In the body of the message, type:

  subscribe framers

Other tools and technologies

Web sites

- www.pdfzone.com

  A web site with resources for Adobe Acrobat users. It includes tips and tricks, resources, and more.

- www.ucc.ie/xml/faq.xml

  The XML FAQ. A resource for those who are new to XML.

- www.wwpusers.com

  An unofficial web site for WebWorks Publisher users. The site includes frequently asked questions, template samples, code snippets, lists of trainers, and more.

- www.winwriters.com

  WinWriters sponsors several conferences focused on providing user assistance and online help.

- www.guru.com

  An excellent site for freelancers of any type.

Mailing lists

- wwp-users

  A mailing list for WebWorks Publisher users. Discussions are usually highly focused on WebWorks Publisher–specific topics. To subscribe, visit:

  groups.yahoo.com/group/wwp-users/

  You will need to register with the web site to sign up for the list, but registration is free.

- Adobe Acrobat

  PDFZone sponsors several PDF lists, including a beginner list, an advanced list, and a list for developers. Get details about the different lists at:

  www.pdfzone.com/discussions/

- Help Authoring Tools & Techniques (HATT)

  A mailing list for help authors. To subscribe, visit:

  groups.yahoo.com/group/HATT/

  You will need to register with the web site to sign up for the list, but registration is free.

- xml-doc

  A mailing list for technical authors interested in XML. To subscribe, visit:

  groups.yahoo.com/group/xml-doc/

You will need to register with the web site to sign up for the list, but registration is free.

Professional organizations

- Society for Technical Communication (STC), www.stc.org

  The STC is a professional association for technical writers and other technical communicators. The STC web site includes information about the organization, its conference, and links to various local chapters. Chapters are mostly in the United States, but there are also several in other parts of the world.

- American Society for Training and Development (ASTD), www.astd.org

  The ASTD is a professional association for trainers and instructional designers.

- Association for Computing Machinery Special Interest Group for Documentation (ACM SIGDOC), www.acm.org/sigdoc

  The ACM SIGDOC is for senior technical communicators.

- Institute of Electrical and Electronics Engineers Professional Communications Society (IEEE PCS), www.ieeepcs.org

  The IEEE PCS helps technical communicators and engineers develop their writing skills. The IEEE PCS sponsors a conference each fall.

- tekom, www.tekom.de

  An organization for technical writers in Germany.

Ergonomics

- Occupational Safety & Health Administration, www.osha.gov

  The web site for the Occupational Safety & Health Administration contains a great deal of information about ergonomics and work-related injuries.

- Typing Injury FAQ, www.tifaq.com

  The Typing Injury FAQ explains repetitive motion injuries related to typing and how to prevent them.

- odp.od.nih.gov/whpp/ergonomics/office.html

  This web page illustrates how to set up a workstation to prevent a variety of conditions including back pain and eyestrain.

Job banks

In addition to the technical writing–specific job banks listed here, most of the major job sites, such as monster.com and dice.com, have technical writing positions.

- STC job banks

  Many STC chapters post job listings on their web sites. To find your local chapter's web site, go to the STC national web site:

  www.stc.org

  The Carolina chapter has a particularly good job bank:

  www.stc-carolina.org/jobs/weeklyjobs.shtml

- About.com

  About.com has many links about technical writing jobs. Go to:

  search.about.com

  and search on "technical writing jobs."

C Tools information

Table 7 lists tools by category and includes information about the companies selling them. This list is not comprehensive; instead, it lists some of the more popular tools.

Table 7: Tools

Type	Software	Vendor
Communications (email)	Eudora	Qualcomm, Inc. www.eudora.com
Communications (FTP)	Fetch (Macintosh)	Fetch Softworks fetchsoftworks.com
	WS_FTP (Windows)	Ipswitch, Inc. www.ipswitch.com
Compression utility	StuffIt (Macintosh)	Aladdin Systems, Inc. www.aladdinsys.com
	WinZip (Windows)	WinZip Computing, Inc. www.winzip.com

Table 7: *Tools (continued)*

Type	Software	Vendor
Content/text development	Epic Editor	Arbortext, Inc. www.arbortext.com
	FrameMaker	Adobe Systems, Inc. www.adobe.com
	Microsoft Word	Microsoft Corp. www.microsoft.com
	XMetaL	Corel Corp. www.corel.com
Desktop publishing	InDesign	Adobe Systems, Inc. www.adobe.com
	PageMaker	Adobe Systems, Inc. www.adobe.com
	QuarkXPress	Quark, Inc. www.quark.com
File conversion and single sourcing	Acrobat	Adobe Systems, Inc. www.adobe.com
	AuthorIT	AuthorIT Software Corporation Ltd. www.authorit.com
	MIF2GO	Omni Systems, Inc. www.omsys.com
	WebWorks Publisher	Quadralay Corp. www.webworks.com

Table 7: *Tools (continued)*

Type	Software	Vendor
Graphics	FreeHand	MacroMedia, Inc. www.macromedia.com
	HiJaak Pro	IMSI www.imsisoft.com
	Illustrator	Adobe Systems, Inc. www.adobe.com
	Paint Shop Pro	Jasc Software, Inc. www.jasc.com
	Photoshop	Adobe Systems, Inc. www.adobe.com
	Snapz Pro (Macintosh)	Ambrosia Software, Inc. www.ambrosiasw.com
	Microsoft Visio	Microsoft Corp. www.microsoft.com
	xv (UNIX)	John Bradley/Trilon, Inc. www.trilon.com/xv/
Help	Doc-To-Help	ComponentOne LLC www.componentone.com
	RoboHelp	eHelp Corp. www.ehelp.com
Project management	Microsoft Project	Microsoft Corp. www.microsoft.com

Table 7: *Tools (continued)*

Type	Software	Vendor
Web authoring	ColdFusion	Macromedia, Inc. www.macromedia.com
	Dreamweaver	Macromedia, Inc. www.macromedia.com
	GoLive	Adobe Systems, Inc. www.adobe.com

Free email providers

There are several web sites that provide free email service, including:

- www.eudoramail.com
- www.hotmail.com

Some web portals, such as Yahoo! (www.yahoo.com), provide free email service.

D   Sample doc plan

The following pages contain a sample doc plan. The sample does contain information about financial matters, but it does not contain specific pricing information.

Documentation plan

Mr. Rip Van Winkle
WinkleWart Technologies
Research Triangle Park, NC 27709

September 15, 2003

WinkleWart Technologies' WinklePlan software allows end users to import data about their mattress manufacturing environment, run simulations of different sleep environments, and output reports. The simulation process helps end users predict how new equipment and other changes will affect the quality of sleep.

The WinklePlan software needs documentation, including a *User's Guide, Quick Start,* and HTML Help. This proposal outlines how Scriptorium Publishing will develop and deliver the documentation.

Audience

The audience for the WinklePlan documentation consists of managers who work in a mattress manufacturing environment. Typically, the software users would be analysts who understand the environment very well and can build an accurate simulation model.

These individuals are typically somewhat computer-literate. They are familiar with Microsoft Windows operations and using a mouse. They are also likely to be familiar with Microsoft Excel.

Deliverables

This section outlines the scope of the project and what services Scriptorium Publishing will provide.

User's Guide

Scriptorium Publishing will develop a task-oriented *User's Guide* for the WinklePlan software. The *WinklePlan User's Guide* will explain the concepts underlying the software and how to use it. The document will contain the following sections:

- Introduction
- Importing data
- Sleeping
- Generating reports

The book will include a table of contents and index.

Scriptorium Publishing will write the manual, edit it, lay it out, index it, and production edit it.

For this guide and the *Quick Start,* we will use the structure and formatting defined by our DocFrame solution, a structured authoring package we developed for use with Adobe FrameMaker document processing software.[1]

Estimated page count: 100 pages

Quick Start

The *WinklePlan Quick Start* will explain how to complete a single project using WinklePlan software. It will provide a specific example so that new users can understand how the software works.

Because this book will be very short, we do not plan to include an index. The book will have a table of contents.

Scriptorium Publishing will write the manual, edit it, lay it out, and production edit it.

Estimated page count: 30 pages

HTML Help

To produce the online help, Scriptorium Publishing will use the DocFrame solution to convert the source material from the *User's Guide* to HTML Help.

Estimated topic count: 80 topics

Deliverable format

The books will be delivered in PostScript or PDF format suitable for professional printing.

The online help will be delivered in HTML Help format.

---

1. See www.scriptorium.com/docframe for more information.

Upon payment of all invoices, WinkleWart Technologies can also request and receive the FrameMaker source files.

Receivables

WinkleWart Technologies will provide the following to Scriptorium Publishing:

- Source files for the current version of the documentation.
- Any feature lists, functional specifications, product design documents, or asset lists that might be helpful in planning the project.
- A working version of the software with a stable, frozen interface. If the interface is not frozen, WinkleWart Technologies will note which portions are least likely to change.
- Ready access to technical experts.
- Timely reviews. Technical experts will return reviewed chapters within three business days.

  Note that a changing interface, difficulty in contacting technical experts, and delays in reviews will lead to change orders, an increase in the total price of this project, and delays in delivery.

Tools

Scriptorium Publishing will use some or all of the following tools:

- FrameMaker—for authoring
- DocFrame—for conversion to HTML Help

- Acrobat—for conversion to PDF
- Paint Shop Pro—for screen captures

Location

Work will be performed at the offices of Scriptorium Publishing.

Deadlines

Exact deadlines will be negotiated between Scriptorium Publishing and WinkleWart Technologies.

Intellectual property

Upon receiving payment in full for all invoices, Scriptorium Publishing will transfer all copyrights for the documentation to WinkleWart Technologies. Scriptorium Publishing retains ownership of any processes developed to complete the project.

Financial

Costs

Based on the information provided (PowerPoint presentation with outline of product), the cost for each deliverable is in the table on the next page.

Note that changes in the product will lead to change orders and a corresponding increase in cost. Furthermore, these estimates are based on the page counts provided in the deliverables section. If the final page count is more than 10 percent higher than the estimates in this proposal, change orders will result.

Fixed-price bid

Deliverable	Cost
User's Guide	$
Quick Start	$
HTML Help	$
Book and HTML Help design (DocFrame)	$
Total	**$$$$**

This is a fixed-price bid.

Any estimates made by Scriptorium Publishing for the cost of any services shall be estimates only. Whenever estimated prices are quoted, Scriptorium Publishing shall use all reasonable efforts to perform the relevant services at the estimated price but in no event shall such estimates constitute a fixed price or not-to-exceed price agreement between the parties in respect of such services.

Whenever the parties agree to a fixed price or a not-to-exceed price, the applicable estimate must expressly so state.

Payment

For the fixed-price portions of the project, Scriptorium Publishing will invoice WinkleWart Technologies for one third of the cost upon approval of this proposal, one third after 30 days, and the final third at the project's conclusion.

For the hourly portions of the project, Scriptorium Publishing will invoice WinkleWart Technologies for hours incurred at regular intervals (weekly or every other week). The invoices will include a breakdown of hours by task (for example, writing, editing, and illustration). Upon request, a breakdown of hours by individual and by day are available.

Invoice terms are net 15, and a charge of 1.5 percent per month will be applied to any overdue account. Overdue accounts that result in a collection by an attorney will be charged for reasonable attorney fees.

Cancellation

Should WinkleWart Technologies cancel this project, Scriptorium Publishing will charge WinkleWart Technologies at $/hour for the work already completed up to the date of notice of cancellation and for work required to terminate the project after notice of cancellation.

Proposal duration

This proposal is valid until September 29, 2003.

Agreement

By signing below, Scriptorium Publishing and Winkle-Wart Technologies agree that the specifications, costs, and terms described in this proposal are acceptable to both parties, and that work may begin.

Scriptorium Publishing Services, Inc.   WinkleWart Technologies

By:_____   By:_____
Sarah S. O'Keefe   Rip Van Winkle

Date: September 15, 2003   Date:_____

Index

Numerics

A

B

You just spotted a discount!

As a reader of **Technical Writing 101**, you get a special discount when you purchase books from Scriptorium Publishing's online store at **store.scriptorium.com**.

To get **10 percent off** a purchase of books and workbooks, type **TW101** as the coupon code when you check out. It's our way to thank you for reading **Technical Writing 101**. And standard shipping in the United States is free!

Discount not valid on FrameScript: A Crash Course, The Masters Series: FrameMaker 6, and scratch-and-dent titles. You can use the coupon code once; if you want to purchase multiple books, put them in one order to get the discount on all eligible titles.

scriptorium.com

You want structure *when?!?*

- ◆ Standards-based single sourcing
- ◆ XML round-tripping
- ◆ Cost-effective
- ◆ Extensible
- ◆ Easy to install
- ◆ XSL transformation
- ◆ Structured FrameMaker files
- ◆ Developer and author documentation
- ◆ Output to print, PDF, XML, HTML, HTML Help, and JavaHelp

Structure in minutes, not months.

www.scriptorium.com/docframe   sales@scriptorium.com
919-481-2701